Connecting the Dots

Connecting the Dots

Environment, Health, Safety, and Sustainability

Kathy A. Seabrook
Global Solutions, Inc.
Mendham, New Jersey, United States

Tanis J. Marquette
Sapheyr, LLC.
North Wales, Pennsylvania, United States

WILEY

Published by John Wiley & Sons, Inc., Hoboken, New Jersey.
Published simultaneously in Canada.

Library of Congress Cataloging-in-Publication Data Applied for

Paperback ISBN: 9781394204236

Cover Design: Wiley
Cover Image: Courtesy of Adam Fox and Woodard & Curran

Set in 9.5/12.5pt STIXTwoText by Straive, Chennai, India

Contents

Foreword

The authors and contributors in *Connecting the Dots: Environmental, Health, Safety and Sustainability* provide the reader with insight into a critical topic for the early twenty-first century. Sustainability, ESG (Environmental, Social, and Governance), and CSR (Corporate Social Responsibility) are subjects that corporate leaders and the public have keen interest to demonstrate their commitment and understanding to protect people and the planet. Responsible businesses can demonstrate their value system by deploying the United Nations Global Compact Ten Principles. These principles provide the structure for company's responsibility in areas of human rights, labor, environment, and anti-corruption. There are influential drivers for businesses to manage their ESG impacts. They include consumer and investor expectations, risk mitigation, cost savings, access to capital, attract talent, and reputation enhancement. Executives and external stakeholders should continue to assess the role of business in solving the social and environmental ills that confront our planet. Ultimately, effective sustainability integration will allow an organization to achieve its full potential and competitive advantage.

The value of health, safety, and human capital and the relationship to the people side of sustainability is a key consideration. When organizations develop leading health and safety practices, employees are more engaged, productive, and loyal. Leading organizations will consider workplace health and safety as material to their stakeholders. The book describes impact pathways that will demonstrate how safety and health processes and other proactive mitigations positively impact and devise value for workers, communities, and businesses. The value created ultimately uplifts society and will improve the global economy.

Furthermore, human capital equates to an educated and skilled workforce that drives technological advancements and innovation which are key components in sustainable development. Education and training investments enhance workforce capabilities and demonstrate social responsibility and mitigate inequalities.

Climate change effects are a growing concern. An increase in global temperatures influences the hydrological cycle and affects permafrost and glaciers. There is an increasing number of severe weather events, including devastating floods and violent storms. Heatwaves will become more extreme, and droughts will last for extended periods and frequency will increase. Environmental climate risk has major implications to worker and community health issues. These environmental and public health issues may have a financial impact on a company's ability to supply products and services. Additionally, proactive risk management is a book highlight that conveys the opportunity for physical and transitional risk reduction and leveraging sustainability-related initiatives.

The book explores critical facets that define responsible business practices and evaluate the concept of materiality which is the framework of decisions shaped by significance. The essence of resilience as an indispensable force amid the dynamic challenges to our global economy is outlined. *Connecting the Dots: Environmental, Health, Safety and Sustainability* will navigate the landscape of disclosures and reporting, unraveling the importance of transparent communication in fostering trust. The reader will peer into the delicate ecosystem of biodiversity and recognize its pivotal role in sustaining life and the interconnectedness of all living beings. Examining the threads of supply chain is analyzed which aids in the understanding of responsible sourcing and production.

Additionally, the perils of greenwashing are outlined, as the book shines a light on the imperative of genuine environmental commitments amidst a sea of deceptive practices. A challenge for ESG professionals is that many businesses prioritize short-term financial gains over long-term sustainability. Quarterly earnings pressure can discourage investment in ESG initiatives with longer payback periods. To foster trust, business must operate to the highest ethical standards. Human capital emerges as a focal point by emphasizing the ethical treatment and development of the workforce as a cornerstone of sustainable business.

Finally, the intricate web of reporting standards and frameworks and how to utilize a structured approach to guide organizations toward accountability is examined. As you proceed through the landmark publication, the reader will gain insight that not only educates but inspires action which will foster a collective commitment to a world where sustainability is not just a goal but a pathway for sustainable communities.

Ultimately, *Connecting the Dots: Environmental, Health, Safety and Sustainability* describes a comprehensive and systematic way to think holistically and act on challenges confronting all interested parties to promote and advance sustainability and ESG.

Darryl C. Hill, PhD, MBA, CSP
Senior Vice President, Safety and Security
First Student, Inc.

Firstly, I thank the authors, Kathy Seabrook and Tanis Marquette, for giving me the opportunity to contribute via this foreword to their book, it is a real honor.

This is a subject that I have dedicated over 15 years of my life to trying to change the business conversation to be more people centric. As I started pulling together my notes, I realized that this was going to be a real challenge in deciding what to leave out!

Sustainability as a word was drawn to my attention in the late 1990s while working in Environment, Health, Safety (EHS) for a global aluminum and packaging company. At the time it seemed to be an extension of Corporate Social Responsibility incorporating business. I remember some company EHS departments adopting the title EHS and Sustainability. Sustainability was summarized as "People, Planet, Profit" and the idea was to drive profitable and sustainable business taking care of the environment and people. I still think this is still valid.

However, since then the world of EHS has been on a roller coaster involving increasing scrutiny from investors and other stakeholders. Unfortunately, this has resulted in more and more focus on the environmental aspect of sustainability to the extent that many people today associate sustainability only with protecting the environment and nature. Rightfully so, you may say, and to some degree I would agree. By protecting the environment and nature we are not only protecting the other species that live on this planet but also ourselves and maybe prolonging our future. We humans are the source of most of the earth's environmental challenges . . . however, we are also the solution.

So, we are right in the center of the sustainability equation and for us to be innovative and creative we must be healthy, safe, and well. This is Maslow's hierarchy of needs in action. We take care of ourselves and our loved ones and make sure they are safe, often before we take care of others and start to protect the environment. It's the same at work as it is at home. How can we expect people to be motivated and passionate about their work, protecting the environment if they fear for their safety and health or those of their loved ones?

According to the International Labor Organization 2.93 million people lose their lives every year due to work-related accidents and illnesses.

The conclusion is evident; a focus solely on the environment and nature and not on people and their health and safety makes no sense.

Stakeholders including investors and rating agencies have come a long way in the past 20 years, and more and more pressure has been put on companies and governments to act to protect the environment. There has been so much focus on environmental management that some companies are highly recognized for their sustainability, even while they continue to have fatalities and serious irreversible injuries and illnesses at work. We carried on working in our silos driven by stakeholders who also had a siloed approach, until the COVID-19 pandemic changed perspectives. Now the world is waking up to the fact that people-centered sustainability is vital to creating a win-win future where people and planet are protected, and businesses thrive. We need to stop working in our silos and help business understand the true value of the people that work for them. Valuing people means taking care of their health, safety, and well-being while at work.

This book has come at the right time, when we need to break the silos and work together. This book unpacks the interdependency of environment, and people health and safety. It gives direction, toward an integrated world, yet will allow you to adapt the trajectory to your needs and culture.

We are slowly coming full circle in the sustainability agenda back to where we started, with sustainability focused on people as well as planet. Successful companies of the future will certainly be those that have an integrated approach to sustainability; those that do not adapt are unlikely to survive.

Putting people at the center of sustainable organizations involves building bridges and seeking collaboration between EHS professionals, other business functions, and external stakeholders to embed human capital principles into decision-making at all levels. Marquette and Seabrook, in this book, unpack what sustainability is, identify the stakeholders, and give practical guidance to help readers determine where they are in their sustainability journey, develop their own roadmaps and, through an integrated approach to sustainability, achieve cultural transformation to create an organization that cares for planet and people as well as profit.

Whether you are new to the subject or have been working in sustainability for years, I would recommend this book. It will help you kick start or boost your organization's sustainability journey by identifying where you are today and where you should be heading tomorrow but above all it will help you put people at the center of this journey.

Enjoy the read #bethehummingbird

Malcom Staves, FIISRM, FIChemE
Global Vice President Health & Safety
L'Oréal

Acknowledgments

To my daughters, Kate and Isabelle, and niece, Erin, amazing women in their own right and without whom I would not understand the importance of family and collectively working toward a better future for all. I love you so very much.

I am very grateful for the many colleagues and friends who have contributed to this book and my sustainability, environmental, health and safety journey over the years. You have inspired me to say "yes" to this hard work, for the benefit of people, workers, communities, society and our planet. Thank you Darryl Hill, Malc Staves, Natalie Nicholles, Mike Wallace, Dave Stangis, Gina Mercurio Brown, Ashok Garlapati, Vic Toy, Jim Howe, David Smith, Sally Swingewood, Nancy Leppink, Zack Mansdorf, Mike Seymour, John Dony, Linda Tapp, Pam Walaski, Fay Feeney, Dee Arp, Martin Cottam, Tom Cecich, Mike Totterdell MBE, Tom Reeves, Mike Oberhill, Kahlilah Guyan, Dr. Georgi Popov, Choke Huckuntod, Brenda Henry, Neil McFarlane, Chet Brandon, Lawrence Sloan, Joanne West, Jan Thomas, Anna Jolly, Olivia Goransson, my co-author, Tanis Marquette and many others.

Thank you to all those who contributed to this book through interviews and knowledge sharing. We are all part of a bright and collective future! #TrueNorth

Kathy A. Seabrook

To Joel, for your unlimited patience, Nutmeg and Ginger for the days upon days of company, Auntie Suzie, for listening even when none of it made sense, Kianna and Caeden for the endless cheerleading, and Sierra for always checking in and keeping me on task. Your love is my everything.

Thank you to Monica and Jim making my words better.

For my colleagues near and far, this is a cumulation of our work together. We are as we have experienced, as those who have touched us. I am so thankful to have been surrounded by such talented leaders.

And to Meagan and Susan. . . my bookends in a library filled with titles of every genre and now one of my own. To you, there are no words a memory can't replace.

Tanis J. Marquette

Glossary

Artificial Intelligence Artificial intelligence, or AI, is a technology that enables computers and machines to simulate human intelligence and problem-solving capabilities (IBM 2024).

Benchmarking Benchmarking is defined as the process of measuring products, services, and processes against those of organizations known to be leaders in one or more aspects of their operations. Benchmarking provides necessary insights to help you understand how your organization compares with similar organizations, even if they are in a different business or have a different group of customers (ASQ 2024).

(The) Capitals Approach In the context of business, a capital is a resource companies can draw upon to operate. These resources can be financial, human, relationship, and natural capital (Howard and Seabrook 2023). A capitals approach enables companies to understand how their success is directly or indirectly underpinned by nature (natural capital), business relationships (social capital) and people working in their company (human capital), which empowers the company to make decisions that offer the greatest value across these capitals (Capitals coalition 2021). See Chapter 3 *Sustainability and ESG Key Concepts* for further discussion and examples of how these capitals benefit a company and its value.

Corporate Social Responsibility (CSR) See Chapter 3 *Sustainability and ESG Key Concepts.*

Dependency Social and human resources and relationships that a business needs in order to create or sustain business value. Resources and relationships include a skilled, engaged workforce, worker health, value chain reliability and resilience, and a rule of law (Capitals Coalition 2019).

Environment, Social, and Governance (ESG) See Chapter 3 *Sustainability and ESG Key Concepts.*

Externality A consequence of an action that affects someone (something) other than the agent (company) undertaking that action, and for which the agent (company) is neither compensated nor penalized. Externalities can be either positive or negative (WBCSD et al. 2011). An example of a negative externality is inadequate hazardous waste disposal impacting local water supplies, resulting in public health issues in adjacent communities. A positive externality is reduced water use in a company's operations increases the availability of fresh water to adjacent communities.

Financial Capital Monetary assets required for a business to provide goods and services. These monetary assets are equity or debt (Investopedia 2024).

Human Capital The benefits a company receives from investing in its people. Human capital value comes in the form of skills, knowledge, commitment, and an individual's energy and drive contributed to their work. It is important to clarify that human capital valuation is not about assigning a monetary value on a person (employee's) life. Valuation is determining the qualitative or quantitative value created by the knowledge, skills, and abilities the workforce contributes to a company (Capitals Coalition 2021). The Organisation for Economic Co-operation and Development (OECD) defines human capital as the "stock of knowledge, skills and other personal characteristics embodied in people that helps them to be productive" (Organisation for Economic Co-operation and Development 2023).

Human Rights Adopted on December 10, 1948 by the United Nations General Assembly, the *Universal Declaration of Human Rights* set out a list of recognized human rights in the wake of World War Two. The *Declaration* defined human rights as rights entitled to everyone, without discrimination, "regardless of race, sex, nationality ethnicity, language, religion or other status" (United Nations 2024). Human rights include freedom from slavery and torture, freedom of opinion and expression, the right to education and to work, and the right to life and liberty. June 2022, the International Labor Organization's 1998 *Declaration on Fundamental Principles and Rights at Work* (1998) to include "a safe and healthy working environment" as a fundamental principle and right at work. (International Labor Organization 2022)

Impact(s) (1) "The effect(s) of organizations' actions on people and the natural environment" (Impact Management Platform 2024). (2) "A positive or negative persistent change in well-being experienced by a person, group of people, nature, a business, etc. that occurs as a result of an activity. Impact examples include access to employment and apprenticeships, consumer trust, inequity, community volunteering" (Capitals Coalition 2016, 2019, 2021).

Impact Driver "A resource that is a measurable non-product output of a business activity." An example of a resource is implementing an ergonomic program or forest regeneration project (Capitals Coalition 2019).

Impact Pathway "The sequence that links organizations' actions with their effects on people and the natural environment." (Capitals Coalition 2019)

Internality "A persistent change in well-being experienced by a person, group of people, nature or the company itself that occurs as a result of an activity. That change and resulting impact can be positive, negative, intended or unintended. When the impact is to the company such as its reputation, competitive advantage or financial stability, it is known as an 'internality'" (Capitals Coalition 2019).

Living Wage "The minimum income necessary for a worker to meet the basic needs of themself and their family, including some discretionary income" (SHIFT 2023).

Materiality Under IFRS, information is material if omitting, misstating, or obscuring, and it could reasonably be expected to influence the decisions that the primary users make on the basis of the financial statements (International Financial Reporting Standards 2024).

Natural Capital The benefits that a company receives from nature and its resources. This includes clean air, clean water, biodiversity, nutrient-rich soil to support crop growth and food sources, and access to raw materials such as wood and precious metals (Capitals Coalition 2016).

Produced Capital The flow of benefits that a company receives from investing in machinery and infrastructure. It is typically communicated in monetary terms (Capitals Coalition 2019).

Resilience Our ability to withstand and recover is called resilience. The true test of resilience is how well we can bounce back. Resilience is societal, economic, and ecological. Understanding risk and preparing today can help protect the things we care about. Communities can mitigate flooding through natural shorelines, and we can work to understand how chemical and biological changes to our ocean impact marine life and habitat. We cannot overrule Mother Nature, but there are actions that we can take together to build resilient communities and support a healthy ocean, sustainable fisheries, and thriving communities and economies (National Oceanic Atmospheric Administration 2024).

Social Capital Benefits that a company receives from the communities in which they operate. These benefits include resources, relationships, trust, and belonging. An example of this occurred during the COVID-19 pandemic. In some communities, this social connection was very strong and kept businesses viable as a result of the strong relationships and social capital those companies had built up with their workers, communities, and customers over time (Capitals Coalition 2021).

Stakeholder A stakeholder is a party that has an interest in the company's success or failure. A stakeholder can affect or be affected by the company's policies and objectives. Stakeholders can either be internal or external. Internal stakeholders have a direct relationship with the company either through employment, ownership, or investment. Examples of internal stakeholders include employees, shareholders, and managers. On the other hand, external stakeholders are parties that do not have a direct relationship with the company but may be affected by the actions of that company. Examples of external stakeholders include suppliers, creditors, and community and public groups (Corporate Finance Institute 2024).

Sustainable Development Development that meets the needs of the present without compromising the ability of future generations to meet their own needs. UN World Commission on Environment and Development (1987) This includes environment, economic and social impacts. See Chapter 3 Sustainability and ESG Key Concepts.

Valuation A representation of value. It is a way of representing something that can influence decision-making.

Value The relative importance of something.

References

ASQ (2024). What is benchmarking? https://asq.org/quality-resources/benchmarking (accessed 30 June 2024).

Capitals Coalition (2016). Natural capital protocol. https://capitalscoalition.org/capitals-approach/natural-capital-protocol/?fwp_filter_tabs=guide_supplement (accessed 10 October 2023).

Capitals Coalition (2019). Social and human capital protocol. https://capitalscoalition.org/capitals-approach/social-human-capital-protocol/ (accessed 8 August 2023).

Capitals Coalition (2021). Human and social capital protocol: a primer for business. https://capitalscoalition.org/wp-content/uploads/2021/12/Social-Human-Capital-Protocol-A-Primer-for-Business.pdf (accessed 5 November 2021).

Corporate Finance Institute (2024). What is a stakeholder? https://corporatefinanceinstitute.com/resources/accounting/stakeholder/ (accessed 30 June 2024).

Howard, A. and Seabrook K.A. (2023). 19 October Interview with Andy Howard, Global Head of Sustainable Investment. Schroders. Howard, Andrew. (interviewee) and Seabrook, Kathy A. (interviewer).

IBM (2024). What is artificial intelligence (AI)? https://www.ibm.com/topics/artificial-intelligence (accessed 30 June 2024).

Impact Management Platform (2024). Impact and the impact pathway. https://impactmanagementplatform.org/impact/ (accessed 25 June 2024).

International Financial Reporting Standards (2024). Materiality in IFRS and financial reporting standards 2 May 2024 https://ifrscommunity.com/knowledge-base/materiality/ (accessed 30 June 2024).

International Labor Organization (2022). A safety and healthy work environment is a fundamental principle and human right. https://www.ilo.org/wcmsp5/groups/public/---ed_dialogue/---lab_admin/documents/publication/wcms:850673.pdf (accessed 5 January 2024).

Investopedia (2024). Financial capital vs. economic capital: what's the difference? https://www.investopedia.com/ask/answers/031715/what-difference-between-financial-capital-and-economic-capital.asp#:~:text=Financial%20capital%20is%20the%20monetary,cushion%20a%20business%20from%20losses (accessed 15 January 2024).

National Oceanic and Atmospheric Administration (2024). What is resilience? https://oceanservice.noaa.gov/facts/resilience.html (accessed 30 June 2024).

Organisation for Economic Co-operation and Development (OECD) (2023). Productivity, human capital and educational policies. https://www.oecd.org/economy/human-capital/ (accessed 10 October 2023).

SHIFT (2023). Accounting for a living wage. https://shiftproject.org/accounting-for-a-living-wage/#:~:text=What%20is%20a%20living%20wage,family%2C%20including%20some%20discretionary%20income (accessed 10 October 2023).

United Nations (1987). "Our common future". Report of the World Commission on Environment and Development. United Nations Digital Library. https://digitallibrary.un.org/record/139811?ln=en (accessed 18 December 2023).

United Nations (2024). Global issues: human rights. https://www.un.org/en/global-issues/human-rights (accessed 5 January 2024).

World Business Council on Sustainable Development (WBCSD), International Union for the Conservation of Nature (IUCN), ERM, PwC (2011). Guide to corporate ecosystem valuation. https://www.wbcsd.org/Archive/Assess-and-Manage-Performance/Resources/Guide-to-Corporate-Ecosystem-Valuation (accessed 10 October 2023).

Section I

Key Concepts

1

Introduction

Welcome to an exciting future. Whether you are a health and safety professional expanding your role into sustainability, or a sustainability practitioner looking to hone your skills, this book is for you.

No matter your level of expertise or the maturity of your company, this book will assist you with practical guidance in strategy development, road mapping, action planning, and operationalizing to help guide your organization through its implementation journey.

The sustainability and environmental, social, and governance (ESG) ecosystem is continually evolving. In this book, we have laid out core concepts in Section I that provide a solid foundation from which to apply the implementation tools that follow in Section II. The Sustainability Framework and Roadmap methodologies provide the structure to "future-proof" you and your organization through effective evergreen strategy development that considers the applicability of appropriate standards and frameworks as well as country-specific regulations, and stakeholder expectations.

By laying out the foundational tenants of sustainability, this book enables you to identify, assess, manage, and disclose your sustainability impacts, risks, opportunities, and most importantly, to imbed this mindset into decision-making across the business, thereby enhancing overall performance and resiliency. It can be used to begin your sustainability journey, as a guide for continual improvement, or anywhere in between.

There are many drivers for a company to begin its sustainability and ESG journey. They include customer demand, investor inquiry, or a decision by the board of directors to embrace the preservation of human and natural resources needed for a viable, resilient, sustainable company. For whatever reason, here you are. So, where to start?

Connecting the Dots: Environment, Health, Safety, and Sustainability,
First Edition. Kathy A. Seabrook and Tanis J. Marquette.

Every company is different and poses unique challenges. This differentiation may be by industry sector, internal and external risks, impacts and dependencies, geographic or geopolitical location, customer requirements, business model, and/or regulatory environment. Therefore, a tailored approach to sustainability is required to be successful.

This book will assist you in navigating sustainability and ESG opportunities as well as the challenges. You will find it packed with real-world examples and insights. We have gleaned the knowledge and expertise from leading experts: Chief Sustainability Officers, VPs of Sustainability, operational leaders, investors, environmental economists, professional organizations, and nonprofit/nongovernmental organizations (NGOs) as well as experts in environmental health and safety (EHS). Through the combination of insights from these subject matter experts, research, and collective work of the authors in sustainability, ESG, and EHS, this book brings a practical approach, strategic framework, implementation road map, and guidance that has led to success for many companies in achieving their sustainability and ESG goals.

We bring together a body of knowledge around sustainability and ESG to inspire practitioners and business leaders alike to seek out the opportunities that abound for their company and to find that seemingly hidden value, across environmental stewardship, social responsibility, and economic contribution. A win-win for all.

Finally, we trust this book proves to be a valuable tool for you and your company to assist with strategy development, action planning, and operationalizing sustainability and ESG. The guidance and new thinking, such as developing impact pathways, can assist you and your company in identifying and managing material issues, impacts to and dependencies on, that may influence the decision-making of investors, lenders, insurers, customers, suppliers, regulators, and your workforce.

Enjoy the read; our hope is you will have the tools needed, a framework and roadmap, to advance your company's sustainability and ESG goals. We wish you well as you begin or continue your sustainability journey.

In this world of sustainability, EHS professionals do not need to seek out a seat at the "table"; we are the "table." EHS is foundational in this ecosystem.

It all starts with the #TrueNorthTest. Are you and your company future fit?

Kathy A. Seabrook and Tanis J. Marquette

2

The Evolution of Sustainability

To provide context, this chapter chronicles key sustainability and occupational health and safety (OHS) sustainability-related milestones since the 1980s. These key sustainability-related milestones are grouped by decade with descriptions below reflecting forward progress on sustainability and ESG, including the intersectionality of human, social, natural, produced, and financial capital management.

An important observation in these key sustainability-related milestones is the generational shift in an evolving sustainability mindset and how it is influencing businesses around the world (Global Solutions, Inc. and Seabrook 2024) Figure 2.1.

In his role as corporate director of health, safety environmental protection, and sustainability at GrafTech International Ltd., Chet Brandon shares his experience over the past 10 years with potential and new employee expectations. "The new, younger generations have been raised in a world where society has expectations that businesses will be ethical, provide positive value beyond profit, keep people safe and enhance our communities. They've come to expect, and are vocal about, that. And when we hire them as employees, they bring that expectation with them." In interviews, Brandon says, "these potential employees want to know the value the company brings to society in addition to profit and enriching stakeholders" (Brandon and Seabrook 2023).

Keep this sustainability-related cultural shift in mind as you follow the evolution of sustainability through the key milestones highlighted in this chapter.

Connecting the Dots: Environment, Health, Safety, and Sustainability,
First Edition. Kathy A. Seabrook and Tanis J. Marquette.

Key Sustainability-Related Milestones

Figure 2.1 Key sustainability-related milestones. Source: Global Solutions, Inc. and Seabrook (2022, 2024).

1980s

World Resources Institute

The World Resources Institute (WRI) was founded in 1982. It is based in Washington, DC with offices in Brazil, China, Colombia, India, Indonesia, and Mexico and regional offices in Africa and Europe.

According to WRI, their focus is on "research-based approaches to meet people's essential needs, protect and restore nature, stabilize the climate and build resilient communities" (World Resources Institute 2024). To do this, they are focused on people, nature, and climate, creating lasting change by embracing the motto of "count it, change it and scale it" (World Resources Institute 2024). This motto translates to research and testing evidence-based approaches in various countries that have large populations of vulnerable people, much of the world's remaining natural landscapes and produce significant greenhouse gas emissions. This geographic focus is key to impact and scaling.

The evidence-based approaches integrate theories of change, targets, and processes for assessing, improving, and scaling progress on impact. For example, transforming the way food and energy are produced and used or designing cities in a more sustainable way impacting people, nature, and climate. This is measurable impact. In order to affect change and scale around the world, WRI partners and engages with other entities such as the World Business Council on Sustainable Development, World Economic Forum, and the Capitals Coalition.

United Nations (UN)

In 1987, the United Nations World Commission on Environment and Development (WCED) published the groundbreaking *Report of the World Commission on Environment and Development: Our Common Future.* This report is also known as the *Brundtland (Commission) Report,* named after former Norwegian Prime Minister Gro Harlem Brundtland, who chaired the WCED at the time (United Nations 1987). In this report, a case was made for sustainable development. Today, the Brundtland report's definition of sustainable development is still relevant, "development that meets the needs of the present without compromising the ability of future generations to meet their own needs" (United Nations 1987).

According to Nitin Desai, a senior adviser to the Brundtland Commission and a key contributor to the report, "the main long-term impact of the report is that we can no longer talk of economic and environmental policy in separate compartments" (UN Commission on Sustainable Development 2007). While the definition of sustainable development seemed to be focused on the environment, it also incorporates the interconnectivity between society, the economy, and sustainable development. The concepts of poverty, lack of equity and environmental justice, both environmentally and socially, are also imbedded in the construct of sustainable development. According to the Brundtland Report, "the sustainability of ecosystems upon which the global economy depends must be guaranteed. And the economic partners must be satisfied that the basis of exchange is equitable." The report further clarifies "when relationships are unequal, rooted in dominance of one over another, there is no basis for interdependence, economic development, and the sustainability of nations" (United Nations 1987). This acknowledges and actively recognizes the value and interconnectivity between environmental policy and protection, economic policy and development, and society as a whole, where society incorporates the social dimension of sustainability.

1990s

World Business Council on Sustainable Development

The WBSCD was founded in 1995 as a global coalition of CEOs from 200+ companies around the world. According to WBCSD, its membership has "annual revenues of USD 7 Trillion and spans more than 35 countries and represents 20 major industrial sectors." Their network also includes national and regional business councils and other organizational partnerships in developing countries.

The purpose of WBCSD is to "advocate for progress on sustainable development . . . in a world where resources are increasingly limited" (GHG Protocol 2024). It does this through various platforms where companies share sustainable development experiences and best practices. They work with governments, non-governmental (NGOs), and other organizations.

WBCSD has been a catalyst for many sustainable development projects such as the establishment of the Greenhouse Gas Protocol and the Capitals Coalition and their social and human capital and Natural capital protocols.

GRI

GRI, formerly known as the Global Reporting Initiative, but now only use their acronym GRI, originated in Boston, United States (US), in 1997 as a collaboration between CERES, the Tellus Institute and the UN Environment Programme. The catalyst for GRI's creation was the 1989 Exxon Valdez incident occurring in the Prince William Sound, Alaska, US in 1989. When the Exxon Valdez oil tanker ran aground, it spilled 11 million gallons of oil and, according to the National Oceanic and Atmospheric Administration (NOAA), resulted in "one of the largest environmental disasters in U.S. history" (National Oceanic and Atmospheric Administration 2024). The intent of GRI was to create an accountability mechanism for responsible environmental, social, economic, and governance issues, which resulted in global disclosure standards for these issues (Global Reporting Initiative 2024).

The first GRI global guidance framework was published in 2000 and in 2016, the first sustainability disclosure standards were published as GRI became a formal standards development body. In 2018, the GRI 403 Occupational Health and Safety disclosure standard was published (Global Reporting Initiative 2024). Today, GRI is the most comprehensive set of disclosure standards for sustainability reporters.

Greenhouse Gas (GHG) Protocol

Founded in 1997, the GHG Protocol's purpose is to provide accounting and reporting standards and tools to achieve a "low emissions economy worldwide" (GHG Protocol 2024). Through a 1997 partnership between the World Resources Institute (based in Washington DC) and World Business Council for Sustainable Development (based in Switzerland), the GHG Protocol was established to develop a globally recognized, standardized method to account for GHG emissions. Companies such as BP and General Motors initially worked with WRI to publish a 1998 report called, "Safe Climate, Sound Business." The report foreshadowed the need for standardized accounting and measurement of GHG emissions. In 2001, the first GHG Protocol Corporate Accounting and Reporting Standard (CARS) was developed and published through a multi-stakeholder standard development process that included business, nongovernment organizations (NGOs), and governments. Contributors included WWF, Pew Center, Shell, Norsk Hydro, Tokyo Electric and others (GHG Protocol 2024).

Today, the GHG Protocol Corporate Accounting and Reporting Standard is globally recognized and underpins corporate GHG reporting programs around the world. This includes the measurement and management of GHG emissions in public and private sector companies and their value chains, industry associations, governments, and NGOs. The GHG Protocol provides guidance on how to account for GHG emissions throughout a company's value chain, including measuring emissions from electricity and other energy purchases (GHG Protocol 2024).

Finally, the CDP recognizes the GHG Protocol standards when calculating and reporting greenhouse gas emissions within their ecosystem.

2000s

United Nations (UN) Global Compact

The Global Compact was launched in 2000 by Kofi Annan, the Secretary-General of the United Nations at the time. Envisioned as a global corporate citizenship, sustainability initiative, the Global Compact established 10 principles in the areas of "human rights, working conditions, the environment and anti-corruption." They are voluntary, not mandatory. The Global Compact seeks to mainstream these principles into business activities such as strategy, policies, processes, and procedures throughout the world. Signatories to the Global Compact are expected to integrate these principles (United Nations Global Compact 2024a).

According to the UN Global Compact Network UK: 2022 Annual Activity Report, there are more than 20,000 companies and 3,500 nonbusiness signatories based in over 162 countries (UN Global Compact Network UK 2023). Companies such as Starbucks, L'Oreal, Bayer AG, the Coca-Cola Company, 3 M, Deloitte, Nike and Microsoft are signatories (United Nations Global Compact 2024b).

Climate Disclosure Project (CDP)

Similar to GRI, CDP was once known by their longer name, the Carbon Disclosure Project; however, once their "project" expanded beyond carbon, they transitioned to the acronym, CDP. CDP was founded in July 2000, providing a global disclosure system for investors, companies, cities, states, and regions to measure and manage their environmental impact. Today, CDP provides a third-party reporting system for use by companies, their suppliers, and investors for disclosure and decision-making. 18,000+ companies (globally) reported into CDP in 2022 (CDP 2023). CDP is moving toward an emphasis on supply chain disclosure, Scope 3 reporting. CDP's premise is "supplier engagement drives action globally" (CDP 2023). For example, CDP's Supply Chain Program works with "280 leaders, representing $6.4 trillion in procurement spend" (CDP 2023). And according to CDP, "at the request of their customers, 16,462 suppliers reported to CDP in 2022" and those "suppliers reported saving 70 metric tons CO_2 due specifically to CDP Supply chain member engagement, this is the equivalent to powering over 9.5 million homes for a year" (CDP 2023).

CDP is currently the leading entity for third-party disclosure on CO_2 emissions, water, and deforestation. As part of their 2025 strategic plan, CDP is rolling out new disclosure requirements for plastics, biodiversity, waste, oceans, land, and food.

ESG: The Origin of the Term

In 2004, the United Nations Global Compact published a report entitled: *Who Cares Wins: Connecting Financial Markets to a Changing World*. This is considered the first mainstreamed use of the term and notion of ESG (The Corporate Governance Institute 2023). The report was an output of an investor led initiative to provide guidance on ESG and encourage business stakeholders to embrace ESG long term (UN Global Compact 2005).

United Nations (UN) Principles of Responsible Investment

With over 5,319 signatories, the UN Principles of Responsible Investment (PRI) was founded in 2006. It is an investor initiative, representing USD 121 trillion of assets under management as of December 2022 (UN PRI 2022). The PRI is a partnership between the investment community and the United Nations Environment Programme – Finance Initiative and UN Global Compact. Its primary objective is to promote ESG investment decision-making through the use of the six PRI voluntary investment principles. The PRI supports its signatories by transitioning the principles into practice, with the goal of developing "a more sustainable global financial system" (UN PRI 2022). These six principles were developed in 2005 by an international group of institutional investors. The signatories to the PRI commit to promoting and incorporating ESG issues into investment decision-making, ownership of policies and practices, seeking disclosure from investment entities, and reporting on activities and progress with implementing the principles (UN PRI 2024).

Accounting for Sustainability

According to Accounting for Sustainability (A4S), it was established by HM King Charles III in 2004, when he was The Prince of Wales. His vision: "to help ensure that we are not battling to meet twenty-first century challenges with, at best, twentieth century decision making and reporting systems" (Accounting for Sustainability 2024). A4S networks catalyze companies and their CFOs, investors, and governments to create systemic change, "recognizing that long term sustainable performance relies on an understanding of the interdependency between financial, social and environmental factors." The aim is to drive change so that "sustainability becomes embedded in organizations' strategy, operations and reporting." According to A4S Executive director, Helen Slinger, "the ultimate goal is sustainable business being business as usual and the role that finance and accounting professionals play in achieving this goal" (Slinger and Seabrook 2023). A4S has created active finance related networks and a Knowledge Hub, including case studies on accounting for sustainability, incorporating social and environmental valuations on a company's balance sheet (www.a4s.org).

2010s

International Integrated Reporting Council (IIRC)

Established in 2010, the IIRC was the driving voice, globally, toward integrated reporting. That is integrating sustainability issues into annual reporting mechanisms. IIRC was merged into the International Sustainability Standards Board (ISSB) in 2022 and no longer exists as a separate entity.

Sustainability Accounting Standards Board (SASB)

The SASB was established in 2011 with its first standards published for use by investors for the US capital market in 2018. SASB merged with the ISSB in 2022, creating an international entity to develop globally recognized sustainability-related disclosure standards, by industry sector, across environmental, social, and governance issues. According to the IFRS Foundation who oversees the ISSB, the aim of ISSB is to set forth the relationship between sustainability related issues and creating financial value. Existing SASB sustainability related standards are recognized by ISSB until new industry-specific IFRS standards are developed (IFRS 2022).

Center for Safety and Health Sustainability (the Center)

In 2011, the Center for Safety and Health Sustainability (the Center) was launched, providing over 100,000 occupational safety and health (OHS) professionals in over 70 countries with a stronger voice in shaping OHS-related sustainability policies. According to the advisory board chair, Zack Mansdorf, the Center advisory board chair at the time, "the Center was a collaboration of safety and health professional organizations from the United States, Canada, and the United Kingdom and the first of its kind to recognize and advocate for the interconnection between sustainability and occupational health and safety" (Mansdorf and Seabrook 2024). Its strategic goals were to provide thought leadership as a global source for emerging insights into the measurement, management, public reporting, and impact of safety and health sustainability. Its role was to influence and engage decision-makers in OHS measurement, management, and public reporting. Its purpose was to prepare business leaders (including safety and health professionals) to understand, measure, manage, and publicly report on the importance of safety and health performance, tying it to good corporate governance, sustainability, and corporate social responsibility (Center for Safety and Health Sustainability 2024).

In 2013, the Center published a groundbreaking white paper titled *Current Practices in Occupational Health & Safety Sustainability Reporting*. The aim was to report on the current state of sustainability reporting on OHS performance. To do this, the Corporate Knights 2011 Top Global 100 Most Sustainable Companies sustainability reports were analyzed. The key finding from the analysis was the need for improved OHS reporting in the context of sustainability reporting. The report outcomes had impact. During a panel discussion at the America's Safest Companies conference with Tom Cecich, chair of the Center and Kathy Seabrook, president American Society of Safety Professionals and advisor and future chair of the Center, the concern highlighted by the white paper and the Corporate

Knights ratings was raised with Tony Heaps, editor in chief of Corporate Knights, who was also on the panel. The discussion influenced him to publicly announce, during the panel discussion, that the number of fatalities would be a rating criteria for determining the top Global 100 Most Sustainable Companies going forward.

The report highlighted the following concerns with the 2011 Corporate Knights ratings: (Center for Safety and Health Sustainability 2013):

- High variability in terms and definitions used to report OHS, making it difficult to use reports to compare OHS performance across organizations. Sustainability reports varied using terms such as "rates of injury," "report-worthy injury or incident," "lost day accidents," and "absenteeism."
- Varying formulas used to determine injury, occupational disease, absentee and lost day rates.
- Less than 10% of these companies reported based upon the recognized Global Reporting Initiative's (GRI) Labor-related indicators, which were the reporting standard at that time. This included the use of the GRI formula for calculating injury rates. There was very low reporting on contract or temporary workers' lost day rate and injury rate.
- A high number of fatalities (10 or more) reported by five organizations. One of the 100 Most Sustainable Companies reported 49 fatalities in a year and another reported 81 fatalities over a three-year period.
- No organization reported on fatal occupational diseases.
- Approximately a third of organizations reported on major topics recommended by the Center, on practices endorsed by the OHS community.

As a result of these report findings, the Center went on to develop and publish globally recognized OHS reporting metrics for sustainability disclosure reporting, entitled *CSHS Best Practice Guide for Occupational Health and Safety in Sustainability Reports.* (Center for Safety and Health Sustainability 2016). This report influenced occupational health and safety in the US Securities and Exchange Commission's modernizing human capital regulations and the Global Reporting Initiative 403 OHS sustainability disclosure standard.

In 2020, the Center became a strategic partner of the newly launched Capitals Coalition. At the time, Kathy A. Seabrook, Chair of the Center explained, "it's increasingly clear that natural, social, human capital and economic issues are fundamentally interconnected, so that's the lens that must be adopted." She continued "people are at the core of every business. If leaders are to succeed in delivering a sustainable and resilient future, those components must be addressed together in an integrated system. There's widespread value in doing that" (American Society of Safety Professionals 2020). Thus, the collaborative work and relationship between the Center and the Capitals Coalition commenced.

The Center was a collaborative voice to put people back into sustainability. This culminated in the 2019 People Sustainability Summit: Putting People back into Sustainability hosted by L'Oreal in Paris and Center board member, Malc Staves, the Global Vice President of Health and Safety at L'Oreal.

When the Center was sunsetted later in 2020 during the COVID-19 pandemic, its vision and thought leadership on human capital, sustainability, ESG, and occupational health and safety transitioned to the Capital Coalition Valuing Human Capital in OHS Project and Project group.

Capitals Coalition

The **Capitals Coalition** was founded in 2020, resulting from a merger between the **Natural Capitals Coalition** (founded in 2012) and the **Social and Human Capital Coalition** (founded in 2018) (Capitals Coalition 2024). The Social and Human Capital Coalition was created by the World Business Council on Sustainable Development. The Capitals Coalition is "a collaboration between leading organizations in business, finance, government, science, accounting and standards, civil society and multi-stakeholder groups" (Capitals Coalition 2024). There are over 450 organizations that are members of the Capitals Coalition with over 20,000 partner and collaborator organizations. See Table 2.1 for a sampling of the 450 Coalition members (Capitals Coalition 2024).

Table 2.1 A sampling of capitals coalition members.

• Kering	• WWF
• Coca Cola	• Credit Suisse
• Target	• S&P Global
• Novartis	• European Investment Bank
• CEMEX	• World Business Council on Sustainable Development
• Natura	• World Resources Institute
• Unilever	• CDP
• Dow	• UN Environmental Programme
• AECOM	• Chartered Institute of Management Accountants
• AkzoNobel	• Deloitte
• Cambridge University	• L'Oreal

Source: Adapted from Capitals Coalition (2024).

At its core, the Capitals Coalition seeks to ensure the value of nature, people, and society, in addition to financial value is identified in order to transform decision-making. This translates to the value of human, social, natural, and produced capital in addition to financial capital. This is "capitals thinking" or a "capitals approach" (Capitals Coalition 2024). See Chapter 6 *The Capitals*; Chapter 13 *Value: Creation, Assessment, Valuation, and Impact Pathways*; and Chapter 16 *Glossary* for more information on this capitals approach.

The Capitals Coalition published two Frameworks that provide companies "with tools to identify, measure and value their impacts and dependencies on natural, social. human and produced capital to inform their decision-making" (Capitals Coalition 2024). The first framework is the Natural Capital Protocol, published in 2016 and focused on the value nature creates for companies (Capitals Coalition 2016) The second is the Social & Human Capital Coalition Protocol, published in 2019 and focused on the value people and business-related relationships create for companies (Capitals Coalition 2019). These protocols provide guidance on incorporating a capitals approach to valuation in order to better understand a company's true value and influence its decision-making. Companies such as Novartis (Novartis AG 2021), Natura (Natura and Co 2022), and Yorkshire Water (UK) (Yorkshire Water 2022) have used and published results using this capitals approach to valuation.

The Capitals Coalition hosts the Valuing Human Capital in Occupational Health and Safety (People Sustainability – OHS) Project and Project Group. See the milestone in this chapter for more information on the purpose and work of this Project and Project group, including publishing the *Human and Social Capital Protocol: A Primer for Business*. This work resulted from the collaboration between the Capitals Coalition, the People Sustainability – OHS Project group, L'Oreal, Novartis, Natura, and ERM (Capitals Coalition 2021).

EU Non-Financial Reporting Directive (NFRD)

The EU Non-Financial Reporting Directive (NFRD), published in 2014, requires EU companies that meet specific scope criteria, to publish a non-financial report on their ESG performance along with their annual management report.

Taskforce on Climate-Related Financial Disclosure (TCFD)

In 2015, the Financial Sustainability Board (FSB) created a taskforce to develop climate-related disclosure standards. This taskforce, known as the Taskforce on Climate-Related Financial Disclosure (TCFD), was chaired by Michael R. Bloomberg, the founder of Bloomberg L.P. as well as the U.N. Special Envoy on Climate Ambition and Solutions. Recognized industry experts from "large banks, insurance companies, asset managers, pension funds, large non-financial companies, accounting and consulting firms, and credit rating agencies" were members of the taskforce (Taskforce on Climate-Related Financial Disclosure 2023).

In 2017, TCFD published disclosure recommendations around a framework of governance, strategy, risk management, metrics, and targets. These recommendations and the framework were created to assist investors and other entities with decisions relevant to how reporting companies view and assess their climate-related risks and opportunities.

The TCFD was also tasked by FSB with monitoring the progress of company use of climate-related disclosures. With the transfer of TCFD work to the International Sustainability Standards Board (ISSB) in 2021, the TCFD was disbanded and their recommendations and monitoring activities were also transferred to ISSB (Taskforce on Climate-Related Financial Disclosure 2023).

United Nations (UN) Sustainable Development Goals (SDGs)

In 2016, as part of the 2030 UN Agenda for Sustainable Development, a 15-year plan was approved by all UN Member states. The plan focused on 17 UN SDGs with 169 integrated targets to achieve those goals by 2030 (United Nations 2024). Figure 2.2 depicts the 17 SDGs.

The SDGs are globally recognized and provide a framework for sustainability and sustainable development that is aligned with the GRI (voluntary) and the European sustainability disclosure standards (mandatory). International human rights are grounded in the UN 2030 Agenda for Sustainable Development and tied to the SDGs (United Nations 2015). The International Standards Organization (ISO) also aligns its standards with the SDGs, and some companies track and report on the alignment of their sustainability initiatives with the SDGs (UN Global Compact 2024a).

Figure 2.2 UN Sustainable Development Goals. Source: Reproduced with permission from the United Nations. The content of this publication has not been approved by the United Nations and does not reflect the views of the United Nations or its officials or Member States (United Nations 2024).

The UN SDGs and World Health Organization (WHO) Social Determinants of Health

According to the WHO, the *Social Determinants of Health* are defined by specific circumstances in which "people are born, grow up, live, work and age, and the systems (e.g. social, political, economic) put in place to deal with illness." These circumstances are shaped by politics, economics, and social policies (World Health Organization 2014). The interconnectivity between the World Health Organization and the UN SDGs is through the WHO's evidence based *Social Determinants of Health and UN SDG 3 Good Health and Wellbeing* goal. Public and occupational health are foundational to the social or people side of sustainability. They impact the social fabric of a society, such as food systems, medical care, education systems, economic opportunity, rule of law, human rights, etc. Therefore, it is important for companies to understand the context in which their sustainability activities impact health outcomes in developed as well as developing countries because all countries have a social gradient that is impacted by these *Social Determinants of Health,* leading to and impacting health outcomes for workers, their families, communities, and society as a whole. The WHO suggests health outcomes are influenced positively and negatively by the following social determinants (World Health Organization 2024):

- Income and social protection
- Education
- Unemployment and job insecurity
- Working life conditions
- Food insecurity
- Housing, basic amenities, and the environment
- Early childhood development
- Social inclusion and nondiscrimination
- Structural conflict
- Access to affordable health services of decent quality

These are the areas of greatest impact identified by the WHO and correlate to where a company's sustainability and social responsibility activities meet the highest need in the area of health at the local, national, regional and global levels.

Human Capital Management – 2017

Founded in 2013, the Human Capital Management Coalition (the Coalition) is a provider of capital, represented by 36 institutional investors (asset owners) with over $9 trillion assets under management (Human Capital Management Coalition 2017).

In 2017, the Coalition petitioned the US Securities and Exchange Commission (SEC) to increase mandatory human capital disclosure for issuers in their annual financial reports, i.e. a company's annual 10 K report. This meant the SEC would need to "modernize" (update) its mandatory human capital disclosure requirements. This rulemaking petition recommended mandatory public disclosure of a company's (issuer's) policies, practices, and performance on human capital management. At that time, the SEC only required a company to publicly disclose the number of their employees (Human Capital Management Coalition 2017).

The Coalition argued human capital management or "talent management is material to company performance and profitability long term." It also contributes to "the creation and protection of long-term shareholder value." Their petition and subsequent comments to the proposed ruling also suggested "a large body of evidence links investments in human capital to better firm performance" (Human Capital Management Coalition 2017, US Securities and Exchange Commission 2019b).

In 2020, the SEC published the modernized Item 101(c) in Regulation S – K regarding the Narrative Description of Business and includes human capital-related disclosures. These final disclosure requirements were principles based (i.e. not a list of required disclosures), within a materiality decision-making framework (US Securities and Exchange Commission 2019a, 2020). This area continues to evolve and at the time of this writing, the SEC is exploring additional changes to human capital disclosure legislation.

International Standards Organization (ISO)

ISO published its first sustainability-related standard *ISO 26000: 2010 Social Responsibility* on November 1, 2010. This original standard was confirmed in 2021 (International Standards Organization 2010, 2021).

Founded in 1947, the International Standards Organization (ISO) is a non-governmental organization (NGO) that develops and publishes internationally recognized voluntary, consensus standards. It has 169 national standards body members from around the world. As an example, the American National Standards Institute (ANSI) is a national standards body member of ISO (International Standards Organization 2024a).

As of this writing, ISO has published (or has in development) ninety (90) sustainability and four (4) ESG related standards. In addition, it has published six (6) human capital-related reporting standards. See Table 2.2 for a nonexhaustive list of ISO sustainability relevant standards (International Standards Organization 2024b).

Table 2.2 Published ISO sustainability-related standards.

- ISO 26000:2010 Guidance on Social Responsibility (confirmed 2021)
- ISO 14001 and related standards. Environmental Management
- ISO 45001 Occupational Safety and Health Management System (confirmed 2022)
- ISO 30414:2018 Human resource management: Guidelines for internal and external human capital reporting
- ISO 21928-2:2023 Sustainability in buildings and civil engineering works. Sustainability indicators
- ISO 21929-1:2011 (confirmed 2022) Sustainability in building construction
- ISO 21931-2:2019 Sustainability in buildings and civil engineering works
- ISO 17989-1:2015 (confirmed 2021) Tractors and machinery for agriculture and forestry-Sustainability
- ISO 20121:2012 (confirmed 2017) Event sustainability management systems

Source: Adapted from International Standards Organization (2024b).

In 2018, the *ISO 45001 Occupational Health and Safety Management System: Requirements with Guidance for Use standard* was published and was subsequently confirmed in 2022. In its "Clause 4.0 Context," it incorporates business initiatives and stakeholder expectations that may impact the management of safety and health. Sustainability issues are now required to be taken into account when determining the organization's context (ISO 2018, 2021). Today, there are ISO Standards across environmental stewardship, social responsibility, and governance, including anti-bribery, risk management, procurement, and numerous other valued topics as shown in Table 2.2.

United States (US) Business Roundtable

As an association of CEOs representing large US-based companies, the US Business Roundtable focuses on sound public policy that promotes a thriving economy. On August 19, 2019, the US Business Roundtable published a statement redefining their policy on the principal purpose of a corporation. 181 CEOs were signatories to this updated policy statement that expanded a company's purpose from only serving their shareholders to also delivering value to their "customers, employees, suppliers and communities in which they operate." This translated to investing in employees, dealing fairly with suppliers, and supporting communities (US Business Roundtable 2019b). This also suggested companies take a leading role and support communities where they serve, specifically mentioning public health and the COVID-19 pandemic as an example (US Business Roundtable 2019a).

This expanded purpose of a corporation is aligned with the foundational notion of sustainability and human, social and natural capital management, ultimately creating value for corporation and its stakeholders.

2020s

Capitals Coalition Valuing Human Capital in Occupational Health and Safety (People Sustainability - OHS) Project and Project Group

The Valuing Human Capital in Occupational Health and Safety (People Sustainability - OHS) Project group ("the Project group") is a Capitals Coalition collaboration. It was founded in 2020 on the premise that "People are at the heart of every thriving business and their ability to contribute is influenced by their health, safety and wellbeing" (Global Solutions, Inc. and

Seabrook 2022) and "at the heart of business and the drive for companies to be more sustainable, are people" (Capitals Coalition 2022, 2024).

The Project group is a collaboration of international safety, health, and sustainability professionals and other OHS stakeholders focused on the value and impact of OHS on companies, people, nature, the economy, and society. Its founders and co-chairs are Kathy A. Seabrook, CEO, Global Solutions, Inc. and former chair, Center for Safety and Health Sustainability and Malcolm Staves, Global Vice President Health and Safety at L'Oréal and former board director, Center for Safety and Health Sustainability. Executive Director, Natalie Nicholles champions the Project group on behalf of the Capitals Coalition (Nicholles and Seabrook 2023).

Project group members have included L'Oréal, Jacobs, Nike, dsm-firmenich, Pernod Ricard, Sompo International, Parques Reunidos, Adient, Global Solutions Inc., the American Industrial Hygiene Association (AIHA), Institute for Advanced Risk Management (IARM), Hexion, GrafTech International Ltd., the Royal Society for the Prevention of Accidents (RoSPA), the National Safety Council (NSC), National Examination Board in Occupational Safety and Health (NEBOSH), European Network of Safety Professional Organizations (ENSHPO), American Society of Safety Professionals (ASSP), Institution of Occupational Health and Safety (IOSH), Health and Safety Professionals - Canada (HSPC), and the International Institute of Risk & Safety Management (IIRSM).

The Project group is focused on assessing the value of occupational health and safety (OHS) as a significant component of human capital management and the people side of sustainability. It is this inflection point, where the knowledge, skills, competencies, and attributes of people, their health, safety, well-being and contribution to decision-making, improved performance, and well-being fundamentally connect human capital to a company's value and valuation.

Dee Arp, Chief Quality Officer, NEBOSH represents NEBOSH on the Project Group. She believes "there is power in re-thinking health and safety. Redefining it for the new generation, where people are at the heart of our work and decisions. This is the teaching I think the new professionals need and common ground with the people side of sustainability" (Arp and Seabrook 2023).

Hallmarks of the Project group are its monthly stakeholder engagement calls, annual People Sustainability Summits and collaborations to create a global OHS, human capital, and people sustainability body of knowledge. More on the Project Group's contributions to the OHS-sustainability body of knowledge are noted below.

Project Group – Contributions to the OHS-Sustainability Body of Knowledge

A) *People Sustainability Summit 2022, 2023, 2024 and Proceedings Reports* hosted and supported by L'Oreal, Jacobs, NEBOSH, RoSPA, Global Solutions, Inc., the IARM, Pinsent Masons, and Protex AI. The Summit reports provide an account of these invite-only, thought leadership events focused on people, workers, OHS, sustainability, and the evolving trends noted in Summit keynotes, panel discussions, and workshops.

B) Capitals Coalition *Social and Human Capital Protocol Primer for Business* was published in 2021. In order to facilitate company implementation, individual Project group members collaborated with the Capitals Coalition to produce a business friendly primer to accompany the original Social and Human Capital Protocol published in 2019.

C) The *Valuing Human Capital in Occupational Health & Safety white paper* was commissioned by L'Oréal and the Capitals Coalition for publication in 2022. It contributes to human capital management by demonstrating the value and correlation of people and their health, safety, and well-being to decision-making and the true value of an organization. The white paper focuses on assessing the valuation of OHS impacts and dependencies, providing companies with resources and data to conduct OHS-human capital assessments. Using an "impact pathways" methodology, OHS – human capital assessment approach and a case study on ergonomic programs, the white paper presents a business case for investing in OHS initiatives, their value to workers and society and how they impact well-being. The case is made for the valuation of impacts such as well-being on workers as a result of implementing OHS initiatives (Capitals Coalition, L'Oreal 2022). Impact pathways methodology goes beyond the traditional impact "valuation assessment of avoided medical costs and lost income for workers" (Capitals Coalition 2024). It can provide the true value of OHS, worker contribution and Human Capital to a company's balance sheet.

D) The *Connecting the Dots: Safety, Health and Environmental Sustainability report* was published in 2022. It was commissioned by L'Oreal in collaboration with the Capitals Coalition and prepared by ERM. The purpose of this work is to highlight the interdependencies that span both the ESG and human capital domains to showcase the interconnections between all realms of ESG and their impacts on our health. It depicts the direct and indirect relationships between nature, workplace, and public health systems and provides an example of these interconnections through the lens of climate change.

One of the co-authors of this book (Kathy Seabrook), served as a US delegate to the ISO TC283/Taskgroup 8, indicated this report was an influential resource for the Taskgroup 8's "seed" document used as a starting point to develop the ISO/PAS 45008, *OSH Risks Arising from Climate Change* standard.

E) *The New Value of Safety and Health in a Changing World* research was published in 2023 by National Safety Council, with financial support from Lloyds Register and in collaboration with the Capitals Coalition Project group and its members. This groundbreaking research work is discussed more fully in the Milestone on the National Safety Council's research below.

International Sustainability Standards Board (ISSB) and International Financial Reporting Standards Foundation (IFRS)

The International Sustainability Standards board (ISSB) was founded in 2021 and operates with oversight by the International Financial Reporting Standards Foundation (International Financial Reporting Standards - Foundation). ISSB is a private sector entity that approves and publishes voluntary sustainability-related disclosure standards for use by the investment community as an international baseline. These standards have been recognized by the G20, International Organization of Securities Commissions (IOSCO), and financial authorities from over 40 countries (IFRS Foundation 2023). Some countries such as Brazil are incorporating the ISSB standards into their mandatory, regulatory framework (International Financial Reporting Standards 2023).

The first IFRS globally recognized sustainability-related disclosure standards were published in 2023 under the direction of the ISSB. They are the IFRS S1 General Requirements for Disclosure of Sustainability-related Financial Information and IFRS S2 Climate-related Disclosures standards (International Financial Reporting Standards 2023).

Of note, in 2022, the International Integrated Reporting Council, Sustainability Accounting Standards Board, and the Taskforce on Climate-Related Disclosure merged with the ISSB. It is the intent of these entities to align and integrate their principles, collective knowledge, experience, standards, guidelines, and framework with the development of globally recognized, consistent, decision useful, comparable sustainability-related ISSB disclosure standards.

IFRS Foundation's ISSB sustainability-related standards have been developed for use across industries sectors and environmental, social, and governance issues. According to the IFRS Foundation, the aim of ISSB is to set forth the relationship between sustainability related issues and creating financial value. The current SASB sector-specific sustainability-related standards are in effect until they are replaced by the new IFRS standards in development ((International Financial Reporting Standards 2022).

In 2022, the ISSB clarified that "a company's ability to deliver value for its investors is linked to the stakeholders with whom it works with and serves, the society it operates in, and the natural resources it draws on." This sustainability linkage can result in a company creating, preserving, or eroding value over time by "addressing the impacts, risks and opportunities that can affect its performance and prospects – to ultimately deliver financial value for investors" (International Financial Reporting Standards 2022).

The National Safety Council New Value of Safety and Health in a Changing World Research

The NSC New Value of Safety in a Changing World research report was published in 2023. At the time, it was groundbreaking work connecting the value of safety and health to a business across impact on health and well-being, economics, environmental, sustainability, resilience, ethics, society, and brand reputation. The research report includes real-world examples of safety and health valuation to enable companies to determine the true value of safety and health as it relates to materiality. These new valuation methodologies demonstrate how safety and health initiatives materially create or erode value for a company in a financial way and captured on the balance sheet. The activation guides in the report assist organizations with working through these valuation methodologies to tailor them to their company, industry sector, risk profile, culture, leadership and safety, health, and sustainability maturity (Dony and Seabrook 2013). See Chapter 13 *Value: Creation, Assessment, Valuation and Impact Pathways* for more information on the research report and valuation.

European Union (EU) Corporate Sustainability Reporting Directive

The EU Corporate Sustainability Reporting Directive (CSRD) was published in 2023 and the Corporate Sustainability Due Diligence Directive was published in 2024. See Chapter 5 *Frameworks* for information on this EU regulatory frameworks.

Conclusion

Sustainability and ESG have come a long way since the 1980s and continue to evolve, specifically in the areas of mandatory reporting requirements, assurance and stakeholder expectations around the world. Going forward, watch for more focus on biodiversity, water use and deforestation in addition to climate change as well as the social and governance side of sustainability.

References

Accounting for Sustainability (A4S) (2024). A4S AIMS. https://www.accountingforsustainability.org/en/about-us/overview.html (accessed 24 June 2024).

American Society of Safety Professionals (2020). CSHS [Center for Safety and Health Sustainability] takes lead role in coalition to advance worker safety and health. *Press Release*. https://www.assp.org/news-and-articles/2020/02/05/cshs-takes-lead-role-in-coalition-to-advance-worker-safety-and-health (assessed 9 January 2024).

Arp, D. and Seabrook, K.A. (2023). 15 September 2023 interview with Dee Arp, Chief Quality Officer, NEBOSH. Arp, Dee (interviewee) and Seabrook, Kathy A. (interviewer).

Brandon, C. and Seabrook, K.A. (2023). 8 September 2023 Interview with Chet Brandon, Corporate Director of Health, Safety & Environmental Protection, GrafTech International Ltd. Brandon, Chet (interviewee) and Seabrook Kathy A. (interviewer).

Capitals Coalition (2016). Natural capital protocol. http://www.naturalcapital-coalition.org/protocol (accessed 11 October 2023).

Capitals Coalition (2019). Social and human capital protocol. https://capitals-coalition.org/capitals-approach/social-human-capital-protocol (accessed 8 August 2023).

Capitals Coalition (2021). Social and human capital protocol: a primer for business. https://capitalscoalition.org/wp-content/uploads/2021/12/Social-Human-Capital-Protocol-A-Primer-for-Business.pdf (accessed 17 August 2023).

Capitals Coalition (2022). Connecting the dots: safety, health and environmental sustainability report. https://capitalscoalition.org/wp-content/uploads/2022/12/Connecting-EHS-and-Sustainability_Final_October22.pdf (accessed 8 January 2024).

Capitals Coalition (2024). Valuing human capital in occupational health and safety. Website overview. https://capitalscoalition.org/publication/valuing-human-capital-in-occupational-health-safety/#:~:text=This%20White%20Paper%2C%20Valuing%20Human,be%20core%20to%20decision%2Dmaking. (accessed 8 January 2024).

Capitals Coalition, L'Oreal (2022). Valuing human capital in occupational health & safety. White Paper. https://capitalscoalition.org/publication/valuing-human-capital-in-occupational-health-safety (accessed 17 August 2023).

CDP (2023). Scoping out: tracking nature across the supply chain. Global supply chain report 2022 (website report summary). https://www.cdp.net/en/research/global-reports/scoping-out-tracking-nature-across-the-supply-chain (accessed 3 January 2024).

Center for Safety and Health Sustainability (2013). Current practices in occupational health & safety sustainability report. https://aiha-assets.sfo2.digitaloceanspaces.com/AIHA/resources/CSHS/CSHS_SustainReport_2013_FinalZ.pdf (accessed 13 December 2023).

Center for Safety and Health Sustainability (2016). CSHS best practice guide for occupational health and safety in sustainability reports. https://aihaassets.sfo2.digitaloceanspaces.com/AIHA/resources/CSHS/CSHS_Best_Practice_Guide_Final.pdf (accessed 13 December 2023).

Center for Safety and Health Sustainability (2024). Our history. https://www.aiha.org/public-resources/center-for-safety-health-sustainability (accessed 7 January 2024).

Dony, J. and Seabrook, K.A. (2013). 23 August 2023 Interview with John Dony, Vice President Workplace Strategy, NSC. Dony, John (interviewee) and Seabrook, Kathy A. (interviewer).

Global Reporting Initiative (2024). Our mission and history. https://www.globalreporting.org/about-gri/mission-history (accessed 3 January 2024).

Global Solutions, Inc. and Seabrook, K.A. (2022). The EHS professionals expanding role in human capital, sustainability & ESG. December 8, 2022 American Society of Safety Engineers Virtual Webinar.

Global Solutions, Inc. and Seabrook, K.A. (2024). *Fundamentals of ESG and Sustainability. Safety Focus.* American Society of Safety Professionals.

Greenhouse Gas Protocol (2024). About us. https://ghgprotocol.org/aboutus#:~:text=The%20first%20edition%20of%20the,from%20throughout%20their%20value%20chains (accessed 3 January 2024).

Human Capital Management Coalition (2017). HCMC in the news: coalition petitions U.S. SEC to require increased human capital disclosure. *Press Release*. https://www.hcmcoalition.org/_files/ugd/ee804c_3f93f1323c6545c4804dd5fd866aff6b.pdf (accessed 6 January 2024).

International Financial Reporting Standards (2022). ISSB (International Sustainability Standards Board) describes the concept of sustainability and its articulation with financial value creation and announces plans to advance work on natural ecosystems and just transition. https://www.ifrs.org/news-and-events/news/2022/12/issb-describes-the-concept-of-sustainability (accessed 28 September 2023).

International Financial Reporting Standards (2023). Brazil adopts ISSB (International Sustainability Standards Board) global baseline, as IFRS Foundation Trustees meet in Latin America. https://www.ifrs.org/news-and-events/news/2023/10/brazil-adopts-issb-global-baseline (accessed 28 December 2023).

International Standards Organization (2010 confirmed 2021). ISO 26000:2010, confirmed 2021 Social responsibility. International Standards Organization.

International Standards Organization (ISO) (2018, confirmed 2022). ISO 45001:2022. Occupational health and safety management systems. Requirements with guidance for use. International Standards Organization.

International Standards Organization (2024a). About us. https://www.iso.org/about-us.html (accessed 7 January 2024).

International Standards Organization (2024b). Standards. https://www.iso.org/standards.html (accessed 7 January 2024).

Mansdorf, Z. and Seabrook, K.A. (2024). 6 February 2024 Interview with Zack Mansdorf, former Advisory Board Chair, Center for Safety and Health Sustainability. Mansdorf, Zack (interviewee) and Seabrook, Kathy A. (interviewer).

National Oceanic and Atmospheric Administration (NOAA) (2024). NOAA assessment, remediation and restoration program. https://darrp.noaa.gov/oil-spills/exxon-valdez#:~:text=What%20Happened%3F,environmental%20disasters%20in%20U.S.%20history (accessed 3 January 2024).

Natura & Co (2022). Natura integrated profit & loss accounting 2021 – technical executive summary and insights. https://api.mziq.com/mzfilemanager/v2/d/9e61d5ff-4641-4ec3-97a5-3595f938bb75/d8f2cae6-7a86-1d24-8100-62ae5871c7fc?origin=2 (accessed 2 January 2024).

Nicholles, N. and Seabrook, K.A. (2023). 22 September 2023 Interview with Natalie Nicholles, Senior Director, Capitals Coalition. Nicholles, Natalie. (interviewee) and Seabrook, Kathy A. (interviewer).

Novartis AG (2021). Novartis in society ESG report 2020: valuing our impact. https://www.reporting.novartis.com/2020/novartis-in-society/our-journey-to-build-trust-with-society/valuing-our-impact.html#:~:text=The%20overall%20social%20impact%20of,February%202021%20on%20novartis.ch (accessed 8 January 2024).

Principles of Responsible Investment (2022). Signatory update: October to December 2022 https://www.unpri.org/download?ac=18057#:~:text=We%20added%20140%20global%20organisations,US%24121trn%20of%20AUM (accessed 8 January 2024).

Principles of Responsible Investment (2024). About us: what are the principles for responsible investment? https://www.unpri.org/about-us/what-are-the-principles-for-responsible-investment (accessed 9 January 2024).

Slinger, H. and Seabrook, K.A. (2023). 25 September 2023 Interview with Helen Slinger, Executive Direct, Accounting for Sustainability. Slinger, Helen (Interviewee) and Seabrook, Kathy A. (Interviewer).

Taskforce on Climate-Related Financial Disclosure (2023). About. https://www.fsb-tcfd.org/about# (accessed 7 January 2024).

The Corporate Governance Institute (2023) What is the history of ESG? https://www.thecorporategovernanceinstitute.com/insights/lexicon/what-is-the-history-of-esg/#:~:text=The%20UN%20makes%20it%20official,to%20embrace%20ESG%20long%2Dterm (accessed 4 January 2024).

United Nations Commission on Sustainable Development (2007). Framing Sustainable Development: The Brundtland Report – 20 Years On. https://www.un.org/esa/sustdev/csd/csd15/media/backgrounder_brundtland.pdf (accessed 10 December 2024).

United Nations (1987). "Our common future". Report of the World Commission on Environment and Development. *United Nations Digital Library*. https://digitallibrary.un.org/record/139811?ln=en (accessed 18 December 2023).

United Nations (2015). Transforming our world: the 2030 agenda for sustainable development. https://sdgs.un.org/2030agenda#:~:text=We%20resolve%2C%20between%20now%20and,protection%20of%20the%20planet%20and (accessed 7 January 2024).

United Nations (2024). 17 goals to transform our world. https://www.un.org/sustainabledevelopment (accessed 14 June 2024).

United Nations Global Compact (2005). Who cares wins: connecting financial markets to a changing world. https://www.unepfi.org/fileadmin/events/2004/stocks/who_cares_wins_global_compact_2004.pdf (assessed 4 January 2024).

United Nations Global Compact (2024a). The ten principles of the UN Global Compact. https://unglobalcompact.org/what-is-gc/mission/principles (accessed 7 January 2023).

United Nations Global Compact (2024b). Our participants. https://unglobalcompact.org/what-is-gc/participants/search?search%5Bkeywords%5D=+Deloitte%2C&button=&search%5Bsort_field%5D=&search%5Bsort_direction%5D=asc&search%5Bper_page%5D=50 (accessed 7 January 2024).

United Nations Global Compact Network UK (2023). UN Global Compact Network UK, 2022 annual activity report. www.unglobalcompact.org.uk/wp-content/uploads/2023/02/UNGC-UK-Annual-report-Final-1.pdf (accessed 7 January 2024).

US Business Roundtable (2019a). One year later. https://purpose.business-roundtable.org (accessed 4 January 2024).

US Business Roundtable (2019b). Statement on purpose of a corporation. https://system.businessroundtable.org/app/uploads/sites/5/2023/02/WSJ_BRT_POC_Ad.pdf (accessed 4 January 2024).

US Securities and Exchange Commission (2019a). Exchange act release No. 86614, "modernization of regulation S-K Items 101, 103, and 105." Aug. 8, 2019. https://www.sec.gov/files/rules/proposed/2019/33-10668.pdf (accessed on 6 January 2024).

US Securities and Exchange Commission (2019b). File No. S7-11-19 October 22, 2019 Human Capital Management Coalition comments on the Securities and Exchange Commission's proposed release, Modernization of Regulation S-K Items 101, 103, and 105 Exchange Act Release No. 86614. https://www.sec.gov/comments/s7-11-19/s71119-6322887-194462.pdf (accessed 6 January 2024)

US Securities and Exchange Commission (2020). FACT SHEET: modernization of regulation S - K items 101, 103, and 105. Aug. 26, 2020. https://www.sec.gov/news/press-release/2020-192 (accessed 6 January 2024).

World Health Organization (2014). Social determinants of health: key concepts. https://www.who.int/news-room/questions-and-answers/item/social-determinants-of-health-key-concepts (accessed 17 April 2023).

World Health Organization (2024). Social determinants of health. https://www.who.int/health-topics/social-determinants-of-health#tab=tab_1 (accessed 17 April 2024).

World Resources Institute (2024). About us. https://www.wri.org/about (accessed 4 January 2024).

Yorkshire Water (2022). Our contribution to Yorkshire: annual impact and public value report, period ending 31 March 2022. https://www.yorkshire-water.com/media/udajx4fd/ocy22-report-final.pdf (accessed 8 January 2022).

3

Sustainability and ESG Key Concepts: Sustainability, ESG, the Capitals, Value, Valuation, Impacts, and Dependencies

Taxonomy or a common language in a given discipline such as sustainability, ESG, and Capitals thinking is essential for consistent communication between a company and its internal and external stakeholders. This chapter expands upon sustainability, ESG, and the capitals-related terms and definitions in the Glossary to enable better understanding, communication, and integration of these concepts into the company's business strategy, action planning, and execution.

Sustainability, Environment, Social, Governance (ESG) and Corporate Social Responsibility (CSR) are sometimes used interchangeably in business and society. While there are overlaps in the three, their scope and intent differ. In the context of this book, we use the term sustainability. It provides the broadest intent and stakeholder groups and incorporates the other two key concepts. The rational follows.

Corporate Social Responsibility (CSR)

Let us start with CSR. "CSR is based upon how a company shows up as a good corporate citizen in the world. It reflects on their reputation," according to John Dony, Vice President Workplace Strategy at the National Safety Council (NSC) (Dony and Seabrook 2013) CSR can result in community projects including philanthropy through direct financial support, participating in community volunteering or leading joint environmental remediation projects. These activities are typically local to a company's corporate head office or operational sites and can be considered a company's social capital. The Capitals section below will provide more information on social capital.

Connecting the Dots: Environment, Health, Safety, and Sustainability,
First Edition. Kathy A. Seabrook and Tanis J. Marquette.

Table 3.1 Sustainability and ESG.

"Sustainability is about making a better world now and for future generations. ESG is measuring and reporting thosew things that make a better world." – Michael Cooke, Chief Sustainability Officer

Source: Cooke and Seabrook (2023)/with permission of Michael Cooke.

Sustainability

Sustainability builds upon CSR by adding a company's engagement with climate change, its carbon footprint, greenhouse gas emissions, how it cares for its people, value chain workers and customers as well as its impact on the economy and company governance mechanisms for transparency and accountability in its decision making.

The World Commission on Environment and Development (WCED), in the Brundtland Report, defined sustainable development as "development that meets the needs of the present without compromising the ability of future generations to meet their own needs" (WCED 1987). This includes the environmental, social, and economic dimension of sustainability.

Michael Cooke, a seasoned Chief Sustainability Officer in US and EU-based multinational companies, articulates the distinction between sustainability and ESG in Table 3.1 (Cooke and Seabrook 2023). This mirrors the notion that sustainability is about impact while ESG is about measurement and reporting.

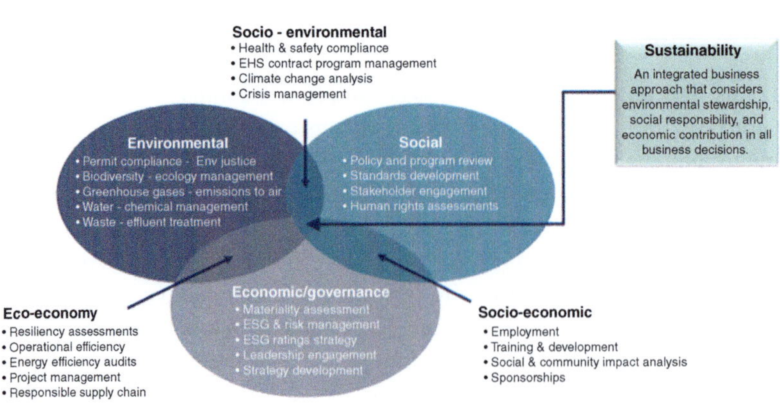

Figure 3.1 Dimensions of sustainability (Marquette 2024).

Table 3.2 Sustainability: the key to success.

"Sustainability is a long term commitment. It is not on a sprint, it's more of a marathon. A successful sustainability professional needs to stay persistent and be patient. This is key to success." – Isidora Diaz Heredia, Global Chief Sustainability and HSE Officer

Source: Diaz Heredia and Seabrook (2023)/with permission of Isidora Diaz Heredia.

Sustainability is focused on longer term impacts a company has on people, relationships, the environment and the economy (externalities) and how people, relationships, the environment and the economy impact the company (internalities) in a material way. Isidora Diaz Heredia is an experienced global chief sustainability and HSE officer. In Diaz Heredia's experience, noted in Table 3.2, sustainability is a long term commitment both for sustainability professionals and their companies (Heredia and Seabrook 2023).

See Figure 3.1 for The dimensions of sustainability-examples. The overlap between sustainability dimensions and ESG issues is synonymous.

At the time of this writing, sustainability continues to be focused on environmental impacts, specifically climate change, biodiversity, and regeneration of natural resources such as water and forests. Some investors and companies who identify and manage these impacts see longer-term value created for their stakeholders.

"Putting people back into the center of sustainability" is a phrase coined by Malcolm Staves, Global VP, Health and Safety at L'Oréal for the 2019 Center for Safety and Health Sustainability *People Sustainability Global Summit: Putting People Back into the Center* of *Sustainability* hosted by L'Oréal (Staves and Seabrook 2023). See Table 3.3; this notion brings to the fore preservation of a company's employees, workers, and their families and communities. It is important to note that people in the context of

Table 3.3 The future of the EHS professional.

"In order to be fit for the future, EHS and sustainability professionals have to evolve, lead and guide organizations to understand sustainability includes people and their capabilities." – Malcolm Staves Global Vice President Health and Safety, L'Oréal.

Source: Staves and Seabrook (2023)/with permission of Malcolm Staves.

sustainability also includes government, supplier, customer, general consumer, and other business relationships where people are decision makers. Environment, health and safety, and sustainability professionals have a role to play in order for companies to be "future fit.".

This social, relationship focus of sustainability is broadening. According to the International Sustainability Standards Board (ISSB), sustainability is "a condition for a company to access over time the resources and relationships needed (such as financial, human, and natural), ensuring their proper preservation, development and regeneration, to achieve its goals" (IFRS 2022). Sustainability is described in the ISSB *General Sustainability-related Disclosures Standard (S1)* as "the ability for a company to sustainably maintain resources and relationships with and manage its dependencies and impacts within its whole business ecosystem over the short, medium and long term" (IFRS 2022). This encompasses the business as a whole and goes beyond an environmental focus. The premise of people sustainability, the social side of sustainability, is still evolving.

Environmental, Social and Governance (ESG)

ESG is a measure of an organization's environment, social, and governance-related risks and opportunities. Figure 3.2 provides examples of ESG-related issues. The concept of ESG was created for the investment community by the investment community. It began to fulfill the need for transparent, reliable and comparable sustainability-related information about a company's material risks and opportunities in order for investors to make more favorable investment decisions.

Dave Stangis is an experienced chief sustainability officer in multinational companies. He views, "ESG factors as useful data. It can be useful information collected from due diligence and risk identification and management perspectives." He considers, "assessing ESG factors as a way to identify future risk and a way to manage and communicate risk, keeping ahead of future risk." From a value creation perspective, Dave continues, "If you know and manage risk, and manage it better than other companies in your industry sector, you are generating competitive value" (Stangis and Seabrook 2023). This underscores value creation through future risk reduction and methodologies to account for competitive value. Opportunities to account for business, customer, and natural environmental value are also presented using the same premise and methodology.

ESG investors have been focused on the environmental side of ESG, specifically on climate change. Consequently, mandatory sustainability

Figure 3.2 ESG issues. Source: Reproduced with permission Global Solutions, Inc / Kathy A. Seabrook.

disclosure standards have followed this focus in many countries around the world. Social and governance issues are beginning to gain recognition. These issues include diversity, equity and inclusion, supply chain integrity, human rights, labor practices, and workplace safety and health. Industry sector, materiality, and a company's experience and reputation in these areas are beginning to influence investors' focus. Stakeholder expectations on how a company is governed are influencing board-level engagement, transparency, assurance and validation, decision-making as well as the integrity of communications (i.e., managing corporate communications, marketing, and potential for "greenwashing").

Lawrence Webb is a globally recognized EHS professional who has worked across-industry sectors. He shares his safety, health and sustainability experience from the various EHS strategy roles he has undertaken. "The larger global brands (and many others) are already on a sustainability journey that includes the safety and health of their people. It is key that they bring their supply chain along with them. If they don't they will only achieve 20% of what's possible because the 80% of work that's being done is by that supply chain, at every level. From a supplier perspective, the motivators are ultimately about what the clients are looking for in sustainability (ESG). The brand of those clients can be damaged by dealing with supply chain organizations who don't look after their people. Webb goes on to say, "It's important for companies to have a value proposition not just for clients, but for their current and future workforce because their people are at the heart of everything they do, and workforces are very transient now." (Webb and Seabrook 2023)

Investors and ESG are twined with publicly held companies. However, it is important to note that ESG can also apply to privately held companies. This is due to customer and other stakeholder expectations. Customers may require ESG-specific information from their value chain, such as a CDP rating. In addition, when a privately held company seeks access to capital for an acquisition or new product line, that debt created maybe subject to the same ESG scrutiny by investors as a publicly traded company. This is dependent upon whether the investment strategy of the lending institution includes ESG-specific disclosure.

The Capitals: Value and Valuation

Human, social, natural, and produced capital, in addition to financial capital, are important key concepts for ESG investors. These capitals create value for a company and its stakeholders and impact investment decisions. Chapter 6 *The Capitals* provides an in-depth discussion and review of these capitals.

Value

In general, value is the relative importance placed on something. Every day, we make decisions based on what we value. For example, the transportation one chooses to get to work is a value-based decision. Choosing to walk, ride a bicycle, take a bus, train, subway, car, airplane, or rideshare is based upon what one values. Typically, this is a balance between travel time and cost. In the context of capitals thinking, one's time could be considered personal or human capital, while the cost of transportation is a direct financial capital consideration. Depending upon what one values, decisions may also be influenced by environmental (natural capital) impact, ease, comfort, and accessibility of the mode of transportation (human and social capital), public health exposure (human and social capital), traffic, the risk of a traffic incident due to inclement weather conditions (impacting human and social capital), the condition of one's car (produced or financial capital), reputation of an airline train company (social capital), etc. These are all personal decisions based on what one values and its relative importance.

Another personal, value-based decision is whether to eat dinner at home or a restaurant. Typical decision-making considerations are socializing with friends, transportation, time, and cost. Choosing healthier food (impacting human, produced, and financial capital) such as organic or farm-to-table, cost between eating out at a restaurant and cooking at home (financial capital), inviting friends to enjoy dinner at your home instead of meeting at a restaurant (social capital.) Eating out at a restaurant may be more expensive (financial capital), but it provides a social event with friends (social capital) as well as no cleanup (human capital). The decision is based on the relative value one places on the options. These value decisions are influenced and traded off the capitals at a personal level. This demonstrates the investment of time and money, and the resulting decisions made, are relative to the value one places on considerations outside of cost (financial capital) alone. For a company, this same notion of value-driven decision-making also applies. See Chapter 6 *The Capitals* for a more in-depth discussion on these capitals.

Another point about value is the alignment of decision-making with the value system of a company. This brings us back to the company's purpose. If a company's #TrueNorth purpose is the context in which decisions are made throughout the organization, the value of people, relationships, and stewardship of natural resources are considered. Ultimately, the company and its decision makers, at all levels of the organizational structure, recognize this alignment with #TrueNorth purpose creates financial, human, social, and natural resource value for both the company and its stakeholders. This is the value proposition.

Valuation

"Valuation is the process of estimating the relative importance, worth, or usefulness of social and human capital to people. Validation can be qualitative, quantitative, monetary or a combination of these. As discussed in the Protocol, the appropriate valuation approach depends on the materiality of the impact or dependency, the available data and needs of stakeholders," according to the Capitals Coalition (Capitals Coalition 2021).

When we speak of value, it is important to understand the difference between value and valuation. Valuation is a representation of value. It is a way of representing something that can influence decision-making and it is typically represented in monetary terms, such as a price for something. Simplistically, the idea of "putting a price on happiness" demonstrates this societal notion. Natalie Nicholles, the Senior Director at the Capitals Coalition, challenges this saying "price is not the right word, it is value"

(Nicholles, Seabrook 2023). Value is more expansive. It enables the ability to represent value through valuation outside of a sole monetary lens. This is where human, social, natural, and produced capitals come into play. They present another way to represent value. The value of people, relationships, and nature.

Let us go back to the example of "putting a price on happiness" and flip it to "putting a value on happiness." How does this change valuation? What is the value one places on happiness or fulfillment in one's work, for example. Monetary valuation (financial capital) is the traditional way to represent or communicate value in the economy. Nicholles goes on to say, "Through trading, through what people will buy and what people will sell, and looking at that behavior, that is one of the most common ways to think about valuation" (Nicholles and Seabrook 2023). Going back to the work example above, fulfillment may be the result of a short commute or remote work options (human capital), a great boss and team (social capital), a company culture that supports people, their development, and families (human and social capital) or alignment on the company's purpose such as a food company promoting healthy food products (human, social, produced and financial capital). In all of these examples, the valuation of fulfillment can be represented or communicated in ways that go beyond direct monetary, financial (capital) valuation.

There is qualitative valuation, where one determines to what extent people think something is important and why, without putting a monetary value on it. Alternatively, there is quantitative valuation, which can also be in terms of monetary or nonmonetary units, but still counts things. For example, how many hours or people it would take to do a task, or the number of life years without a disabling injury (DALY), and the amount of CO2e produced per year. Valuation is important for the value of people, nature, relationships, communities, customers, and suppliers to be visible and tangible in order to be considered in decision-making. To understand valuation is to go beyond the standard view that the economy is only reflected through the lens of direct financial capital (profit and loss). This expanded notion of valuation includes business' impact and dependency on the planet and society in decision-making. That is, decision-making across all stakeholders as well as the company's, including capital markets. It is important to note that if people and the planet are not valued in the context of a company, their value is considered zero, not important, and not reflected in decision-making. For businesses, the economy, and society, this is a missed opportunity.

References

Capitals Coalition (2021). Social and human capital protocol primer. https://capitalscoalition.org/wp-content/uploads/2021/12/Social-Human-Capital-Protocol-A-Primer-for-Business.pdf (accessed 17 August 2023).

Cooke, M. and Seabrook, K.A. (17 May 2023). Interview with Michael Cook, Chief Sustainability Officer. Michael Cooke (interviewee) and Seabrook Kathy A. (Interviewer).

Dony, J. and Seabrook, K.A. (2013). 23 August 2023 Interview with John Dony, Vice President Workplace Strategy, National Safetu Council Dony, John (interviewee) and Seabrook, Kathy A. (interviewer).

Global Solutions, Inc. and Seabrook, K.A. (2024). *Fundamentals of ESG and Sustainability. Safety Focus.* American Society of Safety Professionals.

Heredia, I. and Seabrook, K.A. (19 and 26 June 2023) Interviews with Isidora Diaz Heredia, Chief Sustainability and HSE Officer, Parques Reunidos. Isidora Diaz Heredia (interviewee) and Seabrook Kathy A. (Interviewer).

International Financial Reporting Standards (2022). ISSB (International Sustainability Standards Board) describes the concept of sustainability and its articulation with financial value creation and announces plans to advance work on natural ecosystems and just transition. https://www.ifrs.org/news-and-events/news/2022/12/issb-describes-the-concept-of-sustainability (accessed 28 September 2023).

Marquette, T.J. (2024). Materiality. *National Safety Council Future of EHS Conference.* Louisville, Kentucky (16 February 2024).

Nicholles, N. and Seabrook, K.A. (2023). 22 September 2023 Interview with Natalie Nicholles, Executive Director, Capitals Coalition. Nicholles, N. (interviewee) and Seabrook, Kathy A. (interviewer).

Stangis, D. and Seabrook, K.A. (2023). 18 July 2023 Interview with Dave Stangis, Chief Sustainability Officer. Stangis, Dave (Interviewee) and Seabrook, Kathy A. (Interviewer).

Staves, M. and Seabrook, K.A. (18 May 2023). Interview with Malcolm Staves, FIChemE, Global Vice President Health and Safety, L'Oréal (interviewee) and Seabrook Kathy A. (Interviewer).

WCED (1987). *Our Common Future.* (pg. 43). Oxford: Oxford University Press.

Webb, L. and Seabrook, K.A. (2023). 4 October 2023 interview with Lawrence Webb, EHS strategy director, construction sector. Webb, Lawrence (interviewee) and Seabrook, Kathy A. (interviewer).

4

The Theory of Change: Inputs, Outputs, Outcomes, and Impacts

This chapter provides explanations and examples of **input**, **output**, **outcome, and impact** as they apply to the Theory of Change, The *Sustainability Strategy and Decision-Making Roadmap* (Chapter 9), and *Value: Creation, Assessment, Valuation, and Impact Pathways* (Chapter 13).

The Theory of Change, as depicted in the *Roadmap* (Figure 4.1), enables a shared organizational understanding of how change happens within a company, including the role of a program or action in bringing about change (James 2011). According to the Center for the Theory of Change, a theory of change is "a comprehensive description and illustration of how and why a desired change is expected to happen in a particular context." The *Roadmap* in Figure 4.1 depicts sustainability-related inputs, outputs, outcomes, and ultimately a company's impacts. The goal of a Theory of Change process is to develop information that informs decision making. Identifying company and stakeholder sustainability-related initiatives and their impacts on people, society, the environment, business, and the economy using a Theory of Change process, establishes an opportunity to apply valuation techniques such as value assessment, mapping and an impact pathway methodology. These impacts and value assessment influence decision-making and can translate to advantage over competitors, value created for stakeholders, and improved company performance and market position, leading to better sustainability and overall company performance.

Let us start by reviewing what is meant by **inputs, outputs, outcomes, and impacts** in the context of sustainability, the theory of change, value assessment, and impact pathways.

Connecting the Dots: Environment, Health, Safety, and Sustainability,
First Edition. Kathy A. Seabrook and Tanis J. Marquette.

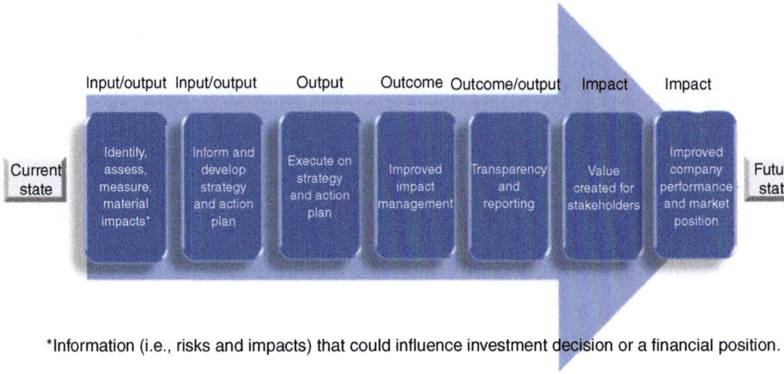

Figure 4.1 The Sustainability strategy and decision-making roadmap (Roadmap). Source: Global Solutions, Inc. and Seabrook (2024).

Table 4.1 Sustainability assessment and implementation tools.

1) Frameworks and Applicability Analysis (Chapter 5)
2) Maturity Analysis (Chapter 9)
3) The Integrated Impacts, Risks, and Opportunities Management Cycle (Chapter 9)
4) The Sustainability (ESG) Strategy Development and Action Planning (Chapter 9)
5) Stakeholder Mapping (Chapter 10)
6) Benchmarking (Chapter 11)
7) Materiality Analysis (Chapter 12)
8) Value Assessment and Mapping (Chapter 13)

Inputs

Inputs are activities or actions that drive change. Table 4.1 provides a list of sustainability assessment and implementation tools that are considered input activities. Let us use the Materiality Analysis tool as an input activity example. As discussed in Chapter 8 *Beginning the Sustainability Journey: Integrated Sustainability Business Model and and Sustainability Culture Change Framework* , Chapter 9 *The Sustainability Strategy and Decision-Making Roadmap*, and Chapter 12 Materiality, the most common way to determine a company's material risks and impacts is to conduct a materiality analysis. A materiality analysis identifies what is important to both the company and its stakeholders. The result is input for developing strategy and an action plan.

A short review of each Sustainability Assessment and Implementation tool found in Table 4.1 is included below. Additional information on these tools can also be found in related chapters, which are referenced where applicable

1) *Frameworks and Applicability Analysis* – This analysis determines the standards, frameworks, and regulations applicable to a company and its sustainability-related activities and reporting (see Chapter 5 *Frameworks*).
2) *Maturity Analysis* – When a company understands its current state of sustainability maturity, it is better able to determine how to advance to an ideal future state of sustainability performance (Accounting for Sustainability 2023). The Seabrook Sustainability (ESG) Maturity Curve provides a benchmark to measure sustainability maturity and is discussed in more depth in Chapter 9 *The Sustainability Strategy and Decision-Making Roadmap.*
3) *Integrated Risk Impacts, Risks, and Opportunities Management Cycle* – A process(es) to identify, prioritize, mitigate, manage, and improve a company's impacts (dependencies), risks and opportunities. It can provide both inputs and outputs in the Theory of Change process. The input informs the development of a sustainability-related strategy and action plan. The output is the actual plan to mitigate risks, impacts and enable the opportunities (see Chapter 9 *The Sustainability Strategy and Decision-Making Roadmap*).
4) *Strategy Development and Action Planning* – The inputs from numbers 1 to 7 above inform a company's current state of sustainability maturity, its material impact risks, stakeholder expectations, and advantage relative to the competition. These inputs inform strategy development and action planning, including prioritization. Strategy directionally sets the company's #TrueNorth, its ideal future state. Action planning prioritizes tangible goals, objectives, and targets, establishes accountability systems and timelines, identifies metrics and key indicators of performance (KPIs), and allocates resources, including capital investment, to achieve the plan. More information on impact pathways is found later in this chapter.
5) *Stakeholder Mapping* – identifies company stakeholders, what is important to them (e.g., sustainability expectations and requirements), and maps sustainability expectations and requirements to the company's priorities. Stakeholders can include key customers, local communities, regulators, suppliers, NGOs, investors, and skilled workers. What is

important differs by stakeholder. Some stakeholders may prioritize human rights, packaging and waste, biodiversity, data use and security, product safety, compliance obligations, public reporting, greenhouse gas emissions, attracting and retaining talent, health and safety, or the composition of a company's board. Others may not. This is why a stakeholder mapping exercise and materiality analysis tools together will provide a process for identifying material topics and legal obligations applicable to the company. More information can be found in Chapter 10 *Stakeholders and Stakeholder Mapping*.

6) *Benchmarking* – This is a strategic comparative-analysis process, typically identifying what needs to be understood or learned by the company doing the benchmarking. This learning can be gleaned from interviews, a literature search, and through disclosure documents to determine a peer group's performance against predetermined criteria. This criteria is based upon a predetermined assessment rubric to assure consistency in benchmark comparisons. The benchmark assessment can be done at many stages in the Sustainability-Related Business Model and Sustainability (ESG) Strategy and Decision-Making Roadmap. This allows for a company's specific needs, sustainability maturity and actions needed for advantage over competitors. See Chapter 11 *Benchmarking*.
7) *Materiality* – Identifies the intersection of material sustainability issues that are important to both a company and its stakeholders. The company determines materiality thresholds and reporting applicability. The output of a materiality analysis provides input for prioritizing, scoping, and managing materiality issues. See Chapter 12 Materiality for a more detailed explanation of a materiality analysis, including definitions, processes, and examples of potential company material risks and impacts.
8) *Value Assessment and Mapping* – Conducting an Impact Pathway can help identify impacts and dependencies on natural, human, social, and produced capital in addition to financial capital (Capitals Coalition 2019). A company's value can be created, enhanced, or eroded depending on its impacts and dependencies. The outcome of an impact pathway and subsequent value mapping process enables a company to identify and track its positive and negative impacts with the intent of enhancing the positive and mitigating the negative impacts. Management of these impacts (and dependencies) on natural, human, social, produced, and financial capital can then be visually mapped to the value being created for the company and its stakeholders. This value map is intended for use by decision makers. More information on impact pathways and value mapping is found later in this chapter.

Outputs

An output is a result of an action taken. Two examples of actions follow.

1) *Executing on Strategy and the Action Plan* – As noted in Figure 4.1, the finalized strategy and action plan are outputs in the *Roadmap*. This output can reflect timelines, quality, extent, and effectiveness of an implementation process. For example, convening a sustainability steering team is an output. The action plan designates the team lead and sets out a requirement for cross-functional department membership, monthly meeting frequency and attendance, documented minutes, internal and external communication, and reporting mechanisms for example. Training delivered for serious injury and fatality prevention processes or on implementing diversity, equity, and inclusion policy as indicated in the action plan is also an output.
2) *Reporting: ESG and Sustainability Performance* – Reporting in general is also an output. What is reported, how transparently, and in what format a company reports can positively or negatively impact its value, competitive advantage, access to capital, as well as stakeholder trust and decision-making. Compliance requirements for annual financial reporting also influence reporting output. Material issues identified by a company and reported in its annual financial report (i.e., US Securities and Exchange Commission [SEC] 10 K) are an example of an output. Additional examples:

   a) The European Union's mandatory Corporate Sustainability Reporting Directive requires companies to report on internal and external material sustainability issues (known as double materiality), while the International Sustainability Standards Board (ISSB) reporting standards, as of this writing, require reporting on internal material sustainability issues (known as single materiality). Note the ISSB standards are voluntary sustainability reporting standards developed for the investment community. See Chapter 5 *Frameworks* for more information on mandatory and voluntary standards and frameworks.
   b) An emerging trend in the uptake of International Sustainability Standards Board (ISSB) sustainability-related reporting standards is beginning to occur in some jurisdictions. In Brazil, the ISSB standards are being incorporated into mandatory regulatory frameworks and phased in over time (International Financial Reporting Standard [IFRS] 2023). As the standards become more globally recognized, other countries such as the United Kingdom, Australia, Canada, China, Japan, New Zealand, Singapore, and Hong Kong are incorporating some requirements of the ISSB IFRS S1 and S2 standards into their mandatory, national regulatory reporting standards (IFRS 2023).

Outcomes

An outcome is the change in status or performance as a result of an input or output as defined above.

Examples of an outcome in the *Sustainability Strategy and Decision Making Roadmap* could be a change in the company's sustainability, ESG, environmental health and safety (EHS), human, social, and natural capital, operational, or commercial performance as a result of an input or output. Input examples include worker safety and health mitigation such as implementing risk assessments,transitioning to tamper-proof packaging product safety design or reducing carbon emissions through a transition to renewable energy. The change in status or performance being worker safety and health risk is better controlled, the risk of product tampering, contamination, and greenhouse gas emissions are reduced. A output from the above inputs is a company annual report that includes these inputs as company Initiatives.

Outcomes can be positive or negative. Here are more examples. A positive change in performance or an outcome may be meeting or exceeding water usage reduction targets or implementing a global program to identify and mitigate human rights issues within the company and its value chain. Another outcome could be an increase in the number of people who attend high-hazard risk or diversity, equity, and inclusion (DEI) training. It is important to note that an outcome is not an impact; outcomes drive impact but are not a measure of impact. Therefore, the number of people attending a high-hazard risk or DEI training is an outcome as a result of an input (the training), but it does not represent a company's impact on employees' safety and health while working in high-hazard-risk areas. Training also does not influence company performance as a result of diverse thought leadership (innovation, new products, services, and market share), nor does training in an of itself raise an employee's quality of life, foster trust in the company, or offer upward mobility, status, or career opportunities.

Impacts and Dependencies

This section of this chapter will guide in better understanding a company's impacts and dependencies, connecting the dots on what they are and how value and valuation apply.

Impacts

Let us begin with impacts. In the broadest context, an impact is the result of an activity that has a significant effect on a situation, entity, or person (Capitals Coalition 2016, 2019). The Organisation for Economic Co-operation and Development (OECD) defines an impact as "a change in people's well-being or the condition of the natural environment that can be attributed to an organisation" (OECD 2023). These definitions are also aligned with the Impact Management Platform's impact definition, "the effect(s) of organisations' actions on people and the natural environment" (Impact Management Platform 2024). Organizations can be companies and their value chains, for-profit, not-for-profit, governmental, or nongovernmental entities. They can represent any industry sector such as investment, manufacturing, construction, technology, automotive, pharmaceutical, energy, utilities, insurance, and other financial types of institutions. Figures 4.2 and 4.3 and Tables 4.2–4.4 include a definition and overarching principles of impact, while Figures 4.2 and 4.4, and Table 4.5 covers the same for dependencies.

Figure 4.2 and Table 4.2 further depict the impact of internalities and externalities. Internalities are activities that have an impact on the company. Externalities are activities that have an impact on an entity outside the company, including its stakeholders, society, the economy, and the natural environment. Connecting the dots, internalities are similar to a single materiality focus: material risk that impacts the company. Double materiality, on

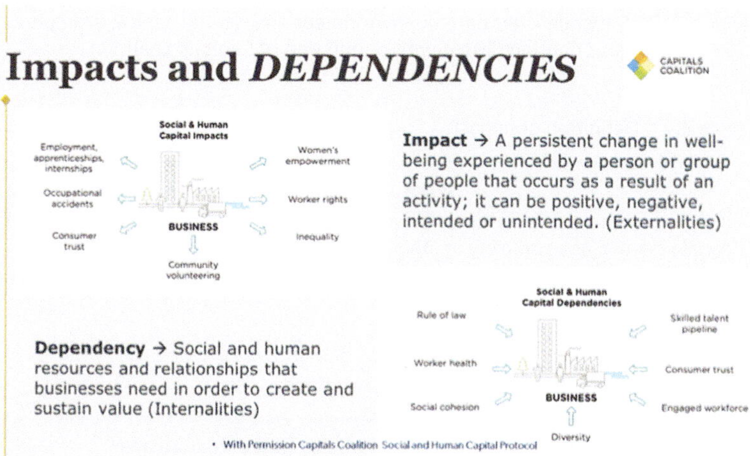

Figure 4.2 Impacts and dependencies: human and social capital example. Source: Capitals Coalition (2019)/with permission of Capitals Coalition.

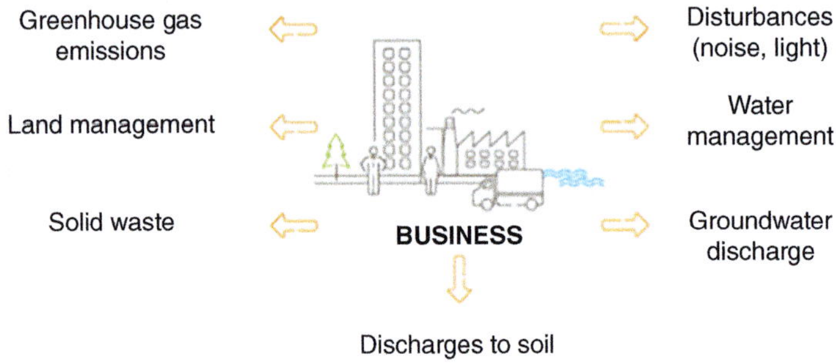

Figure 4.3 Impacts: natural environment and natural capital examples. Source: Capitals Coalition (2016)/with permission of Capitals Coalition.

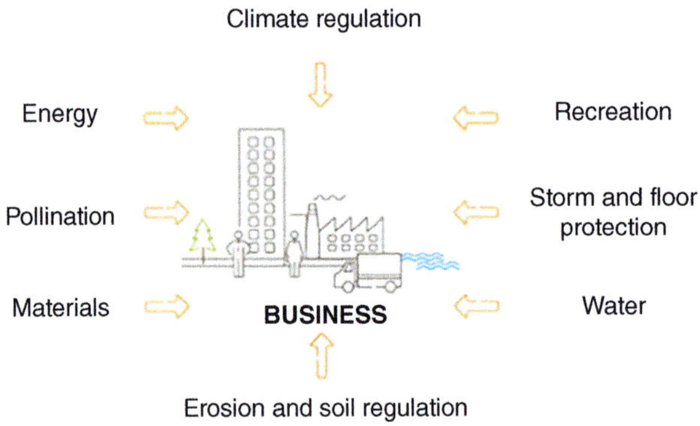

Figure 4.4 Dependencies: natural environment and natural capital examples. Source: Capitals Coalition (2016)/with permission of Capitals Coalition.

Table 4.2 Definition of impact.

What is an impact?
An impact is a persistent change in well-being experienced by a person, group of people, nature, or a company or other business entity that occurs as a result of an activity. That change or impact can be positive, negative, intended, or unintended.

Source: Capitals Coalition (2016, 2019).

Table 4.3 Externalities and internalities defined.

What are impact internalities and externalities?
• Internalities: impact on the company's well-being – its reputation, competitive advantage or financial stability. • Externalities: impact on the well-being of stakeholders, society, the economy, and natural environment.

Source: Capitals Coalition (2016, 2019).

Table 4.4 Differentiating between input, output, outcome, and impact: a DEI training example.

Impact example: DEI training
Providing Diversity, Equity, and Inclusion (DEI) training for operational leaders is an **Input** that drives change and generates an **output,** trained workers. But training in and of itself is not an **impact**. DEI training, for example, can be an **outcome,** where operational leaders with formal and informal influence in the company are proactively identified and trained. Again, this is not an impact. An **impact** of the DEI training is identifying, promoting capable, diverse employees who innovate company practices and create financially viable new products and services. This impacts the company's financial position, its innovation, competitive advantage, and reputation. The employees promoted are also impacted. Their well-being and quality of life are improved with enhanced socioeconomic status, company recognition, career advancement, increased wages, and influence.

Table 4.5 Definition of dependency.

Dependency definition
Social, human, natural resources, and relationships that businesses need in order to thrive, create, and sustain value.

Source: Capitals Coalition (2016, 2019, 2022).

the other hand, can encompasses both internalities and externalities; that is, impacts on the company and the impact the company has on the well-being of stakeholders, society, the economy, and natural environment.

Company impacts can be positive or negative. They can affect employees, workers in value and supply chains, local communities, customers, the economy, society, the natural environment, etc. Positive impacts provide benefits to people, nature, and society. Alternatively, negative impacts impose a burden or cost on people, nature, and society (Capitals Coalition 2019). Impacts are a result of activities or actions, also known as impact drivers. These impact drivers are also considered inputs or outputs that produce outcomes (change), which result in impacts to people, the natural environment, and society (Capitals Coalition 2019; Impact Management Platform 2024). See Table 4.6 for stakeholder impact examples.

For additional clarity on differentiating between input, output, outcome, and impact, the diversity, equity, and inclusion (DEI) training example in Table 4.4 may also be helpful.

Table 4.6 Stakeholder – impact examples.

Activity, action, output, outcome	Stakeholders and impacts
Destruction of biodiversity due to water pollution from site operations	People, employees and other workers, families, communities, society, the economy, and the company, its reputation, market advantage, customers, and suppliers Society (public health with no future biodiversity research and development opportunities)
Access to natural resources due to resource depletion from company deforestation and water use	People, employees and other workers, families, communities, society, the economy, and the company, its reputation, market advantage, customers, and suppliers
Access to skilled labor	People, employees and other workers, families, communities, society, the economy, and the company, its reputation, market advantage, customers, and suppliers
Plant air emissions creating community and worker health issues	People, employees and other workers, families, communities, society, the economy, and the company, its reputation, market advantage, customers, and suppliers
Health, safety, and well-being of employees and value chain workers (physical and psychological health, safety, paid a living wage)	People, employees and other workers, families, communities, society, the economy, and the company, its reputation, market advantage, customers, and suppliers
Improved worker and community health outcomes due to access to company-provided community health-care services	People, employees and other workers, families, communities, society, the economy, and the company, its reputation, market advantage, customers, and suppliers
Improved community public health and well-being due to access to fresh water, clean air, and uncontaminated soil	People, employees and other workers, families, communities, society, the economy, and the company, its reputation, market advantage, customers, and suppliers
Improved supplier relations, business continuity, and resilience due to the company's fair purchase and payment agreements	People, employees and other workers, families, communities, society, the economy, and the company, its reputation, market advantage, customers, and suppliers
Improved employee quality of life with access to upward mobility, status, and career opportunities.	People, employees and other workers, families, communities, society, the economy, and the company, its reputation, market advantage, customers, and suppliers

Activity, action, output, outcome	Stakeholders and impacts
Improved worker well-being due to company identifying and mitigating worker and human rights violations in its operations and value chain	People, employees and other workers, families, communities, society, the economy, and the company, its reputation, market advantage, customers, and suppliers

Source: Adapted from Stangis and Valvoda Smith (2017)/with permission of Emerald Publishing Limited.

Dependencies

Equally important to understand is a company's dependencies necessary to operate. Refer to Table 4.6 for a working definition and examples of a company dependency in the context of impact and value creation. Most companies depend upon their employees and other workers, equipment, energy, raw materials, and suppliers for product or service production. These are the company's social, human, natural, produced, and financial capital. Table 4.7 provides examples of entities, infrastructure, tools, policies, natural resources, etc. upon which a company depends.

Table 4.7 Dependency examples.

Examples: A company's dependency upon:
• Its employees • its value chain workers • its interns, volunteers, onsite contractors • access to skilled labor • its value chain suppliers (e.g., a food company on its supplier's crops) • clean air and water from local water authority • energy • Pollination • its employees • customer trust and loyalty • the economy (access to capital for equipment, infrastructure, or an acquisition) • community (access to suppliers, tax breaks, and permit approvals) • rule of law • information technology • government and regulators (providing a license to operate)

Source: Capitals Coalition (2016, 2019).

There are some companies that refer to their employees as their most important asset. This was seen during the COVID-19 pandemic when essential workers were identified and relied upon to keep business running during the public health crisis.

Companies also depend on the natural environment to generate their products and services. This is its natural capital. An agricultural company is dependent upon good soil health, biodiversity, and land. A hospitality company in a resort area is dependent upon clean air to breathe and fresh water to drink in order to attract customers. An EV battery manufacturer is reliant on lithium and other metals for battery production. Other dependencies on raw materials could be wood, electricity, and biodiverse organisms used in food (i.e., bread and cheese) and pharmaceutical research, development, and drug manufacturing.

Well-Being in the Context of Impact

The concept of well-being is being introduced globally through the evolving context of impact. According to the Impact Management Forum, "the concept of well-being is an umbrella construct that captures the various subjects of impact concerning people. These subjects are also reflected in the legally grounded rights-based approach, in which people's well-being is viewed through the lens of a minimum threshold of indivisible human rights." These rights are reflected in the UN Sustainable Development Goals and the OECD Well-being Framework. The OECD Well-being Framework incorporates 11 dimensions of current well-being and four resources for future well-being also known as "capitals" (OECD 2024).

The progress toward positive impact is the change measured and communicated with stakeholders. In the the example of water usage reduction targets discussed in the *Outcome* heading earlier in this chapter, the actual water reduction achieved is an outcome. One of the positive impacts of this outcome is increased access to fresh water, enhancing employee, worker, and community health and well-being. Risk of dehydration and or the potential of toxic substance ingestion due to a contaminated water source is also reduced. Access to water means better nutrition is enhanced because of increased, cost-effective, locally grown food sources. Additionally, the company is a beneficiary of water security and a reduction in water and operational wastewater treatment costs.

Risks and Opportunities

Identifying impacts and dependencies is akin to identifying risks and potential opportunities, foundational to good risk management. This is an important point in discussions with business leaders because they are familiar with this terminology. They may be more inclined to understand why integrating sustainability and human, social, natural, and produced capital initiatives is an opportunity to reduce risk by influencing culture, strategy, commitment, ownership, and decision-making. As an example, the risk of operational disruptions due to severe weather events presents an opportunity for business continuity and resilience pre-planning to mitigate the outcome of the severe weather event.

In the context of capital investment, an impact analysis is sometimes conducted postinvestment. This happens with due diligence and risk assessments as well. Conducting an impact analysis postinvestment is a missed opportunity to create a better working environment and value for stakeholders. A proactive analysis identifies business-critical risk and opportunities at present and into the future. This includes business continuity and resilience opportunities in the event of a disruption. Tables 4.5 and 4.7 provide additional impact and dependency risks and opportunities to consider.

On the people side, skilled workers needed in the manufacture of a company's products and delivery of services could also be considered a dependency. Assessing these dependencies presents an opportunity for preplanning to mitigate business continuity risks such as labor and supply chain shortages that were experienced during and post the COVID-19 pandemic.

Activities, Impacts, and Dependencies

It is important to note that impacts are the result of a company's activities, not the activity itself. Table 4.8 highlights the importance of measuring impact. "Measuring outcomes and impact, not activities should be a company's focus," according to Dave Stangis, a globally recognized chief sustainability officer. He goes on to say, "there are a lot of company activities, many that are tied to goals and targets such as the number of people trained, how many hours are spent through volunteer programs for the local food bank or how many products are donated. These activities are all good, but they do not measure impact" (Stangis and Seabrook 2023).

When companies rely on inputs, outputs, and outcomes to tell the story, they tend to count activities. For example, a company invested five million dollars, euros, yuan, pounds sterling or yen in the local community, they provided 60,000 hours of training or reached 3,500 students through their equal opportunity internship program. These activities have the potential to drive change and result in impact, but they are not a measure of impact (Stangis and Seabrook 2023). Impact is a more important and relevant measure of how these activities are actually moving the needle on better health outcomes and impacting the quality of life for people, workers, families, and communities.

Table 4.8 Measuring impact versus activity.

"Many philanthropic programs today measure activity – they count how many teachers are trained, how many kids are fed, how many trees are planted. That is good, but what is the outcome of those activities? Those are impact indicators and what I believe are more important to measure." – Dave Stangis, Chief Sustainability Officer.

Source: Stangis and Seabrook (2023)/with permission of Dave Stangis.

Impact can be measured. Interns, employees' well-being, innovation, brand reputation, customer loyalty, employee retention, supply chain disruption, access to capital, and future talent can all be measured by a company. In the National Safety Council's, *The New Value of Safety and Health in a Changing World – Activation Guide and Methodology* case study, Nike assessed how improving its supplier safety practices would impact its business continuity and brand reputation value. Twenty-one suppliers were analyzed during the study, incorporating 71,062 workers (NSC 2023a, b). According to the case study, Nike's brand reputation value was estimated at $USD 6,650,243 and business continuity at $51,173 per year. The annual career and income opportunity was valued at $48,661 and the annual quality of life and avoided household costs were $1,787 and $20,192, respectively. These represent a measure of social and human capital (NSC 2023a, b). While those numbers are not as high, they begin to offer a window into measurable value that was once intangible but can now be incorporated into decision-making. In the Nike case study, decision-making on current and future safety and health interventions and other practices. (NSC 2023b) See Chapter 13 *Value: Creation, Assessment, Valuation, and Impact Pathways* for additional examples of this impact valuation approach.

Once a company understands its impact, the next level in sustainability maturity is to determine how to drive more positive impact, and how ithat impact is aligned to the company's brand (Stangis and Seabrook 2023). An example is a food manufacturer aligning its brand to nutrition and community health within the areas they operate. A car manufacturer transitioning to electric vehicles or a technology company providing communications solutions to enhance equity and inclusion for remote access in developing countries are other examples.

When viewing sustainability and impact through the lens of a company's brand, the sustainability and EHS professional is positioned as a reputation manager. Brand reputation is a lever for influencing the adoption of sustainability and other impact-related initiatives that can help build the brand. This leverage applies to all aspects of the brand, including people, talent,

Table 4.9 Questions to get to impact.

- How did investment in a local community create change? Who is impacted by the change? What value was created for them, other stakeholders, and the business as a result of the change?
- What result did the training and development yield? Are there upskilling and career advancement opportunities? More equitable job promotions? Better socioeconomic positioning and quality of life for workers and communities? Is there a pipeline of highly skilled, capable workers created for the company?
- How did the students working in the internship programs create positive outcomes for the company, customers, suppliers, community, etc.? Are there upskilling and future career opportunities and improved quality of life for the interns, their families and communities?
- Is the company using established, reputable frameworks and standards when creating social measures or key performance indicators (KPIs)? Are the KPIs consistent, comparable, and relevant? Do they measure impact, not activities and inputs?

product, and ethical considerations (Stangis and Seabrook 2023). This next level of sustainability maturity, brand-impact integration can only occur in companies with leaders and other decision makers who understand this sustainability-brand connection.

Table 4.9 explores questions that can get beyond measuring activities, using an inquiry process that identifies impacts that can be measured. These questions are a compilation from the author's experience and interviews with subject matter experts.

To better understand impact, consider whether the company activity moves the needle and creates change and value for people, families, communities, the natural environment, and the company in a meaningful way. This value created by the change is the impact to be measured. Value assessment is a valuation tool to measure the value and impact of an activity that drives change. Impact pathways are a type of valuation tool. Value assessment is further discussed in Chapter 13 *Value: Creation, Assessment, Valuation, and Impact Pathways.*

Connecting the dots between impacts, dependencies, and value, severe weather-related events such as increased temperatures, hurricanes, cyclones, and tsunamis can also increase the likelihood of fires, high-temperature working environments, droughts, flooding, and wind and water damage. When operations are disrupted due to these severe weather events, it can have a negative impact on a company's supply chain dependency, access to labor, resilience, all impacting value.

Additional Impact and Dependency Examples

Examples of impact can be directly or indirectly attributed to company activities such as increased air pollution impacting health outcomes of communities, workers and families, unfair supplier payment practices impacting supplier survival and the local economy or poor working conditions resulting in employee or value chain worker injuries and ill health that impact their well-being, that of their familiy, and community. Reputational and brand integrity, exposure to future litigation and regulatory change, loss of a license to operate, limitations on access to financial, human, and natural resources inhibiting the production of a product or delivery of a service can be indirectly attributed to company activities which impact stakeholder decision-making.

The following are examples of impact and impact metrics.

1) *Living Wage* – A local community's economy can be impacted by how much a company pays its employees, where it buys raw materials and equipment, its supplier purchasing and payment practices and the disposal of its waste and hazardous materials.

   Using the company supplier purchasing and payment practices example above, what are the potential impacts on the local community's economy? First, unfair supplier payment practices could interrupt a supplier's cash flow and ability to operate. When this happens, the supplier will go out of business. When a supplier goes out of business, this can impact the local community economy by reducing local job availability and purchasing capacity to support the community and other local businesses. For the company, when a supplier goes out of business, it can interrupt the company's value chain and ability to provide products and services due to restricted access to the natural resources, component parts or services provided by the supplier. Ultimately, when a local economy cannot support its community, skilled workers move in search of job opportunities, limiting the availability of skilled workers for the company.
2) *Natural Resource Availability* – Companies can impact and be dependent upon natural resource availability. Deforestation, overextraction of minerals, and water usage can directly impact natural resource availability to produce products or deliver on services. Companies can both contribute to shortages of these natural resources, while also being dependent upon them. This poses an impact and risk to its natural and social capitals as well as disruption of the company's business operations, reducing its produced and financial capitals.

3) *Macro Influences* – There have been significant macro influences in the world in recent years. They have included the COVID-19 pandemic, a focus on climate risk and social justice issues, which have highlighted the value, impact, and dependency on people and the natural environment. The COVID-19 pandemic resulted in companies identifying essential workers to assist with the continuity of business across industry sectors. Climate risk has countries around the world focused on carbon emissions reduction targets and energy transition. Companies, investors, and insurers are looking to reduce their climate risk exposures. Insurers, for example, are moving away from insuring homes and businesses in fire and flood-prone areas in the United States such as Florida and California. Finally, social justice issues have companies, governments and academic institutions, along with their stakeholders, reviewing their policies on diversity, equity, inclusion, and pay equity.
4) *Worker Volunteering* – Worker volunteering is another example. To begin to understand its impact, a company needs to identify the change that occurs through worker volunteering; this is the impact to be measured. As mentioned in the *Activities, Impacts and Dependencies* section above, companies typically measure activities, inputs, or outputs (i.e., number of workers volunteering at a local, company-sponsored soup kitchen or homeless shelter or how many pounds of food were donated to a local food bank).

Measuring impact goes beyond activities, inputs, and outputs. To get to the real impact and value created, Dave Stangis suggests, one must continue to ask questions such as 'why is the company doing the activity; what ultimate outcomes can be driven and recreated?' (Stangis and Seabrook 2023).

Sample impact metrics for worker volunteering:

- *Employees* – the impact of volunteering on workers may be enhanced leadership or organizational skill development as well as personal fulfillment and well-being by helping others.
- *Community* – volunteering assures resources for all in the community such as food security and shelter for those in need.
- *The Company* – when workers volunteer, they feel more engaged. When they are engaged, they tend to like where they work. When they like where they work, they stay with their employer. Employee/worker retention is the actual impact that creates value for the company (Stangis and Seabrook 2023).

- *Investors* – Employee retention is a measure of a company's human capital management and if material, employee/worker retention is included in annual reports. This can impact investor decision-making and assessments of corporate value (US Securities and Exchange Commission 2020).

Measurement

The examples of impact and impact metrics above demonstrate that a change in a social, human, natural, produced, or financial capital can result in an impact and value creation or erosion to stakeholders. Impact metrics enable a company to account for the value of positive impacts, resulting in better, more informed decisions and sustainable performance over time.

An improvement in natural capital can result from increased availability and quality of fresh water in a local region. Increased availability of quality water positively impacts workers, their families, the local community, the company, and its suppliers. This impact can also be measured. Measuring the impact of improvements in biodiversity due to changes in a company's activities is also an impact metric. Specifically, a company can measure how its role in improving soil health supports local food production that positively impacts health outcomes and the lives of workers, families, and communities. In both examples above, the company's activities influence its social capital by creating trust through stronger worker and community relationships creating opportunities for the company and all of its stakeholders.

In the worker volunteering example in number four (4) above, value is created for stakeholders beyond the company. Specifically, the change resulting from volunteering impacts and creates value for:

- workers whose sense of well-being is enhanced by volunteering,
- the community that benefits from the employees volunteering, and
- the people for whom the employees provide volunteer services and make difference.

Table 4.10 differentiates between examples of company activity, impact, and dependency metrics that can create value for all stakeholders. Again, working across departments, where appropriate, a company specific monetary value can be developed and assigned to these impacts and communicated to stakeholders.

If we continue to connect the dots, impact can drive measurable, sustainable improvement over time. A positive impact means making things better and reducing risk. It also means being able to measure impact improvements and their associated value to a company and its stakeholders.

Table 4.10 Examples of impact metrics.

Examples Company activity measurement, impact, and dependency		
Activity measurement	**Company impact upon**	**Company dependency upon**
Reduce water consumption in company operations by 50% by 20XX	Local community: • increased water availability • increased access to fresh water in adjacent communities • Public health, food production, improved nutrition Employees and local value chain Workers: • Health, hydration, well-being, trust, loyalty, capacity Natural environment: • Biodiversity	• communities – license to operate • employees, workers • value chain workers • nature/water availability for employee and worker health and operations
Cut greenhouse gases emissions by 50% per product by 20XX	Community: • Improved air quality and public health Employees and local value chain workers: • Improved health and well-being • Access to jobs Company: • Reduced severe weather events • Business continuity	• Traditional and renewable energy sources • Other related natural resources to produce products and services
Increase employee empowerment and engagement to 100% by 20XX	• Well-being for employees, families, communities: • Employee trust and loyalty • Employee retention • Better decision-making	• Workers • Worker participation • Trust, loyalty • Innovation • Productivity • Customer service

(Continued)

Table 4.10 (Continued)

Examples Company activity measurement, impact, and dependency		
Activity measurement	**Company impact upon**	**Company dependency upon**
Implement a 45-day net supplier payment agreement to support supplier continuity by year-end 20XX	Value chain suppliers: • Value and supply chain cash flow, continuity, and resilience Workers: • Job availability	• Value and supply chain continuity for manufacturing products and delivering services
Cut hunger and reduce childhood obesity in the company's local community by 50% by increasing the food product portfolio to the healthiest level	• Community/public health • Employee and worker health • Health-care system • Insurance systems and cost (i.e. medical, company benefits, disability, and life insurance)	• Healthy employees and workers • Communities
Build a new health-care facility in the local community by 20XX	• Improved employee, worker, family, and community health and well-being	• Community trust and loyalty • Healthy employees and workers
Recognize and comply with worker human rights across the company and value chain by 20XX	Employee and value chain workers: • Health and safety • well-being • Fair and equitable pay	• Employees and value chain workers • Value chain

Measuring these tangible impacts on society, the natural environment, and a company's people can be accounted for in a company's profit and loss (P&L) report. Yorkshire Water and Natura are two companies that have used this capitals approach in their P&L reporting (Natura and Co 2022; Yorkshire Water 2023). See Chapter 13 *Value: Creation, Assessment, Valuation, and Impact Pathways* for additional information and examples of the capitals approach to an integrated impact assessment and valuation that can be incorporated in the company P&L.

Competitive Advantage: Business Continuity, Resilience, and Regeneration

Business Continuity and Resilience

When an incident occurs, a company's immediate need is to "weather the storm," both figuratively and literally in the case of a severe weather-related event. The immediate business need is to survive the incident, such as a hurricane, earthquake, fire, explosion, cyclone, flooding, etc., and continue to operate or get the business back up and running. If an infectious disease event or hurricane, cyclone, or tsunami occurs a business must be able to initially withstand the shock.

What about longer-term supply chain disruptions or labor shortages? While businesses want to be able to withstand and recover from these initial shocks, they want to be resilient over time. Sustainability looks at these disruptions through a longer-term lens, in addition to the acute recovery needed from the physicality of a incident such as a hurricane. This longer-term view of resiliency encourages a company to proactively identify and mitigate single-sourced raw materials or components, worker access to the workplace, and test the business continuity plan to assure alternate sourcing capacity - all before an incident occurs.

Regeneration and the Capitals

Company assets, its human, social, natural, produced, and financial capital, can be depleted. When this happens, a company does not receive the flow of benefits from these assets (capitals), which can pose a fundamental risk to the viability and long-term sustainability of the company. Suppliers, customers, raw materials, and people are all considered essential company assets.

Longer-term thinking suggests companies actively plan for business continuity relative to human, social, natural, and produced capitals so they are not depleted. This is where regenerating assets, the capitals, becomes even more important. For example, a company can regenerate human capital by providing employee and other workers access to mental health care and addressing the increasing levels of mental health issues identified in its workforce before, during and after an incident. Creating apprenticeship or internship programs provide a future pipeline of talent for the company. And reforestation assures the availability of locally grown wood for manufacturing and the community are additional examples.

A company and its stakeholder benefit from water regeneration and availability. By eliminating toxic raw materials in the manufacturing process, a company can reduce the risk of pollution and compromising the local water supply. Employees, other workers, and the local community experience better health, while a company reduces its dependency risk on a health-compromised workforce, assuring the supply of a skilled workforce into the future.

A focused approach to preserving the capitals (human, social, natural, produced, and financial) means they will not be depleted, increasing company viability over time. Investment in the foundation upon which the capitals can withstand shocks and other challenges such as water scarcity, public and worker health and safety issues or weather, flood, fire, and pollution events can help to regenerate these capitals. This requires proactive risk management, circular thinking, and business continuity planning. Mitigating risks and adaptation to labor shortages and operational, logistical, value chain, and raw material disruptions requires long-term risk reduction solutions. These solutions incorporate the principles of a circular economy, which include eliminating waste and pollution, reusing, refurbishing, or recycling products, by-products and waste, and regenerating nature. Regenerating nature positively impacts biodiversity as well as availability and company access to natural resources. This circular thinking brings us back to a company's decision-making on investment in proactively strengthening all the capitals. This long term resiliency creates value for the company in the eyes of stakeholders.

Investment in strengthening human and social capital is a focus of the people side of sustainability, regeneration, business continuity, and resilience. When people, employees, workers, talent, teammates, or associates feel cared for, trust is built, and they are more likely to be engaged and willing to support their employers. "Discretionary energy" or extra effort a worker is willing to contribute to their company is what Paul O'Neill, former CEO of Alcoa referred to during his tenure at Alcoa. This extra effort is above the expectations of the job and job description (BC Worker Safety 2014).

Employees and workers tend to help their company "weather the storm" during foreseen, unforeseen and fortuitous challenges. For example, designing a water-saving manufacturing process in a water-scarce area is a foreseen challenge, whereas hurricanes, earthquake, unscheduled downtime event, and construction material delivery delays may be unforeseen or fortuitous. Employees and other workers may choose to go above the expectations of the job and contribute their "discretionary energy" to tackle challenges, problem solve, and work overtime toward longer-term, sustainable business and operational solutions.

Uncertainty, change, and a constant push for workers and teams to make decisions and solve problems creatively are interconnected. Companies that cultivate organizational resilience, whether driven by opportunity or crisis, are able to gain an advantage over competitors (McKinsey and Company 2023). This advantage is fueled by the company's relationship with its people, the company's human, and social capital.

Proactively managing employee and other worker's safety and health can also lead to organizational resilience, specifically in the event of serious injury, illness, or fatalities (SIIF). Resilience incorporates capacity and strength of defenses (operational controls) through preventative actions. This means companies proactively assessing *potential* outcomes of incidents before they occur, versus reactively assessing the *actual* outcomes during a postincident investigation (Dony and Seabrook 2023). A focus on SIIF potential versus SIIF actual. Assessing proactively reduces the risk of the SIIF occurring, while providing an organizational learning opportunity on present and future risk identification and reduction. This strengthens the capacity of defenses and reduces workplace safety and health risk.

Good impact risk management, regeneration, and resilience are markers of a more proactive, learning organization with capacity, strong defenses in place, and leading indicators to enable leaders to make better, more proactive, decisions to enhance human and social capital, sustainability, and EHS performance. In the event of a failure or significant incident, this proactive approach allows a company to fail safe and in a healthy way, resulting in increased organizational resilience. This can create a competitive advantage and mitigate potential material risk for a company, while creating value for stakeholders.

Chapter Summary

In this chapter, we have explored the Theory of Change and how it applies to EHS, sustainability, a company's impacts and dependencies, and EHS performance. We connected the dots on impact risk and dependencies and the importance of good risk management, business continuity, human and social capital, regeneration, and resilience.

Informed decision-making relative to positive sustainability-related internal and external impacts and dependencies is the overarching objective of the *Sustainability Strategy and Decision-Making Roadmap,* Theory of Change, and value assessment. This chapter differentiated between value, activities, and impact, while also demonstrating the interconnectivity of inputs, outputs, outcomes, and impact. Examples throughout this chapter highlighted the importance of the capitals approach, and how it ultimately creates value for the company and its stakeholders.

Chapter 13 *Value: Creation, Assessment, Valuation, and Impact Pathways* provides valuation methodologies that incorporate the Theory of Change, impact pathways, the capitals, impact risk mitigation, and how these activities and processes influence internal and external decision-making.

References

Accounting for Sustainability (A4S) (2023). Maturity maps from the A4s essential guides series. https://www.accountingforsustainability.org/en/knowledge-hub/guides/maturity-maps-essential-guides.html (accessed 20 December 2023).

British Columbia (BC) Worker Safety (2014). Tough decisions: Fareed Zakaria: CNN GPS special. https://www.youtube.com/watch?v=56a3-Sc65M8 (accessed 31 March 2024).

Capitals Coalition (2016). Natural capital protocol. www.naturalcapitalcoalition.org/protocol (accessed 11 October 2023).

Capitals Coalition (2019). Social and human capital protocol. https://capitalscoalition.org/capitals-approach/social-human-capital-protocol/ (accessed 8 August 2023).

Capitals Coalition (2022). Valuing nature and people to inform business decision-making. Coursera Online Course. https://capitalscoalition.org/new-online-business-training-course-launched-today/ (accessed 23 September 2023).

Dony, J. and Seabrook, K.A. (2013). 23 August 2023 Interview with John Dony, Vice President Workplace Strategy, NSC. Dony, John (interviewee) and Seabrook, Kathy A. (interviewer).

International Financial Reporting Standards (2023). Brazil adopts ISSB (International Sustainability Standards Board) global baseline, as IFRS Foundation Trustees meet in Latin America. https://www.ifrs.org/news-and-events/news/2023/10/brazil-adopts-issb-global-baseline/ (accessed 28 December 2023).

James, C. (2011). Comic Relief and Theory of Change. Comic Relief, Jun.

McKinsey & Company (2023). How to implement transformations for long term impact. https://www.mckinsey.com/capabilities/implementation/our-insights/how-to-implement-transformations-for-long-term-impact (accessed 17 February 2024)

National Safety Council (NSC) (2023a). The new value of safety and health in a changing world. https://www.nsc.org/getmedia/b96b7c20-994c-4959-9a36-a8507a2be7e7/nvos-report.pdf (accessed 23 September 2023).

National Safety Council (NSC) (2023b). *The New Value of Safety and Health in a Changing World – Activation Guide and Methodology*. Itasca, Illinois: National Safety Council.

Natura & Co (2022). Natura integrated profit & loss accounting 2021 – technical executive summary and insights. https://api.mziq.com/mzfilemanager/v2/d/9e61d5ff-4641-4ec3-97a5-3595f938bb75/d8f2cae6-7a86-1d24-8100-62ae5871c7fc?origin=2 (accessed 2 January 2024).

OECD (2023). *Policy Guide on Social Impact Measurement for the Social and Solidarity Economy, Local Economic and Employment Development (LEED)*. Paris: OECD Publishing https://doi.org/10.1787/270c7194-en. (accessed 15 February 2024).

OECD (2024). The OECD framework for measuring well-being and progress. Measuring Well-being and Progress: Well-being Research. https://www.oecd.org/wise/measuring-well-being-and-progress.htm (accessed 15 February 2024).

Global Solutions, Inc. and Seabrook, K.A. (2024). Fundamentals class ASSP Safety Focus, 22 Feb 2024.

Stangis, D. and Seabrook, K.A. (2023). 18 July 2023 Interview with Dave Stangis, Chief Sustainability Officer. Stangis, Dave (Interviewee) and Seabrook, Kathy A. (Interviewer).

Stangis, D. and Valvoda Smith, K. (2017). *The Executive's Guide to 21st Century Corporate Citizenship*. Bingley, UK: Emerald Publishing Limited.

The Impact Management Forum (2024). Impact and the impact pathway. https://impactmanagementplatform.org/impact/#:~:text=Definition,people%20and%20the%20natural%20environment (accessed 15 February 2024).

US Securities and Exchange Commission (2020). FACT SHEET: modernization of regulation S-K items 101, 103, and 105, 26 Aug 2020. https://www.sec.gov/news/press-release/2020-192 (accessed 6 January 2024).

Yorkshire Water (2023). Our Contribution to Yorkshire: 2023. *Annual impact and public value report*. https://www.yorkshirewater.com/media/bhxetnfm/ocy-main-report-fy23.pdf (accessed 29 March 2024).

5

Frameworks

Introduction

As introduced in Chapter 2 *The Evolution of Sustainability*, framework is the broad term used to describe structured criteria for defining, implementing, or disclosing sustainability information and performance data. Frameworks take many forms and can be voluntary or mandatory. The scope of the framework criteria can be industry-wide, topic-specific, or cover a broad range of topics that apply to organizations in general. The purpose of most frameworks in use today is to standardize information that is shared and/or analyzed.

What Is It?

Frameworks arise from a need to establish guidelines towards a more sustainable future Sustainability Knowledge Group (2018). From the chapter on the evolution of sustainability, you are aware the landscape of sustainability frameworks has continued to evolve since its inception. In this chapter, we will consider how the various types of frameworks inform organizational initiatives, policy development, and ultimate disclosure strategy.

Frameworks can be divided into six types: Initiatives, Principles and Declarations, Proprietary Frameworks, Consensus Standards, ESG Ratings and Indices, and Regulations. Each type serves to outline sustainability performance criteria to meet the framework's goal or stated purpose. The framework can be issue-specific, for example, Net Zero Initiative whose goal is to "contribute fairly to the only carbon neutrality target that makes sense: net zero at a global scale," or broadly encompassing sustainability

Connecting the Dots: Environment, Health, Safety, and Sustainability,
First Edition. Kathy A. Seabrook and Tanis J. Marquette.

Initiatives*	Principles and declarations	Proprietary framework	Consensus standard	ESG Ratings & indices	Regulation
• Initiatives created to drive performance improvement in spite of lagging regulatory drivers • CDP (formerly known as Carbon Disclosure Project), Science-based Targets Initiative (SBTi), Net Zero Initiative (NZI)	• Principles are a set of aspirational requirements setforth by an organization, typically sector-specific • Principles for Responsible Investment (PRI), UN Global Compact, UN Blueprint for Business, Universal Declaration of Human Rights, UN Guiding Principles	• Disclosure framework that may stem from principles and/or consensus standards, ESG Ratings priorities, or regulation typically focused on an industry or sector • Responsible Care® Sustainability, Ecovadis (supply chain sustainability), USGBC LEED, BREEAM, GRESB	• Non-mandatory, usually consensus criteria for disclosing sustainability performance developed by industry professionals • ISO 45001, GRI 403, Human Capitals Protocol	• Survey-driven evaluations of sustainability performance for the financial industry (primarily investors) • MSCI, ISS, State Street, SustainAlytics, Dow Jones Sustainability Index	• Promulgated regulatory requirement to disclose certain sustainability criteria. May be topic-specific or broad. • US SEC S-K (Human Capital), Sustainable Finance Disclosures Regulation (SFDR),EU Corporate Sustainability Reporting Directive (CSRD)

*Initiatives may ultimately convert to consensus standards over time, e.g. CDP.

Figure 5.1 Framework types.

across all elements of E, S, and G like the ESG Ratings questionnaires offered by the ESG Raters (Net Zero Initiative 2024). The table above provides a description of each framework and examples therein (see Figure 5.1).

We will start with the regulatory frameworks because they are the most concrete, at least typically. For our environmental, health, and safety colleagues, especially those that yield from the jurisprudence side of the house, we know that seemingly black-and-white text is not always straightforward, but we will do our best to make it so!

Regulatory

Regulatory frameworks may be promulgated through different processes based on the country of origin. They represent a requirement to implement or disclose differing aspects of sustainability. For example, the UK Modern Slavery Act (2015) legislation is specific to the offences related to engaging in slavery, servitude, and forced or compulsory labor, whereas the European Union (EU) Corporate Sustainability Reporting Directive (CSRD) outlines requirements for disclosure across broader sustainability performance. CSRD follows requirements of the European Sustainability Reporting Standards (ESRS) as advised by the European Financial Reporting Advisory Group (EFRAG), which are an advancement from the original 2014 Non-financial Reporting Directive requirements originally published covering environmental matters, some human capital, and governance disclosures (see Figure 5.2).

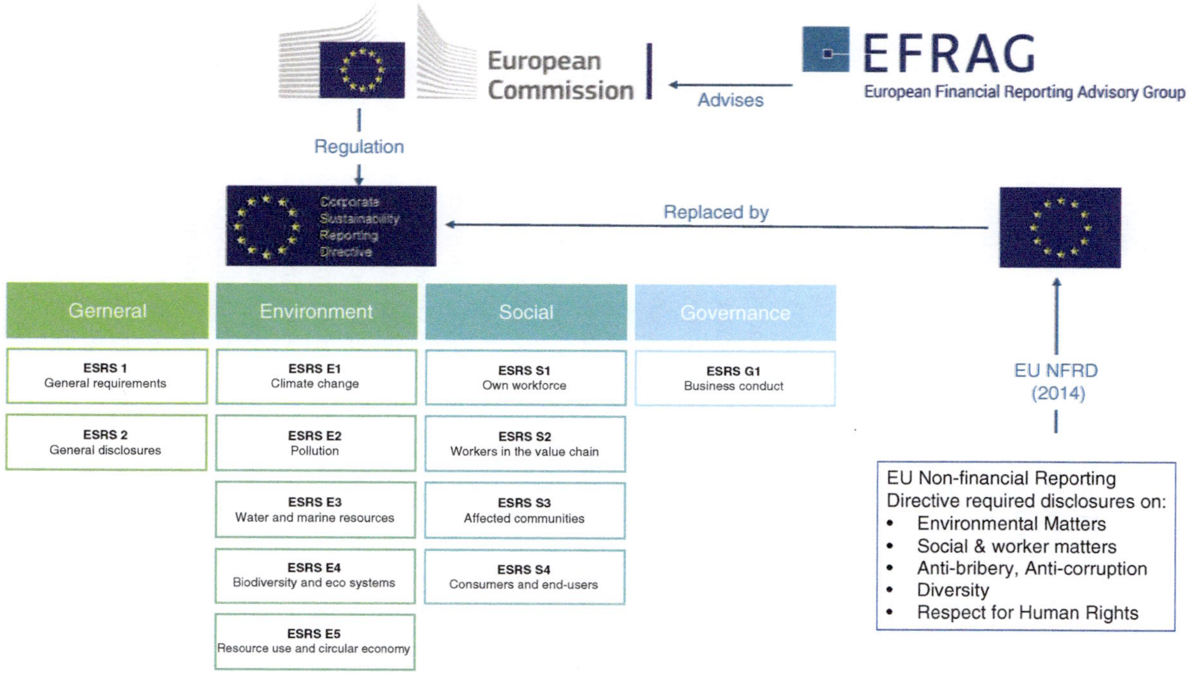

Figure 5.2 Relationship between EU reporting frameworks.

ESG indices are a unique category that provides disclosure information on an investment. Its audience is the investor or financial community. ESG Raters and ratings seek to capture the risk analysis in the investment across environmental stewardship, social responsibility, and governance and assign a value. Each of the global raters has their own unique methodology for creating the weighting algorithm. These frameworks have become instrumental in elevating the conversation globally, so much so that in 2023, the EU promulgated a new directive – the EU Taxonomy proposing new rules for ESG rating providers which will increase transparency for sustainable investment and support the transition and contribution to the Green New Deal in Europe (EC 2023).

Consensus Standards

However, not all frameworks are regulatory in nature. Internationally, the most recognized set of voluntary frameworks is the International Standards Organization (ISO) consensus standards. Consensus standards are developed by a team or teams of experts representing their country's vote. The ISO standard is a culmination of the expert groups' consensus on all elements, including scope, definitions, and content. Each country team works independently on their drafts and come together both virtually and for in-person meetings to follow the structured standards process to develop an initial or revised standards or guidance documents. The ISO organization maintains some of the oldest sustainability frameworks still in use today. However, not all voluntary frameworks are consensus standards (ISO 2023).

The Global Reporting Initiative's framework of disclosure guidelines has evolved over their more than 30 years of existence. The GRI guidelines grew from a series of guidelines that organizations from all sectors and dozens of countries followed into the standards that are in use today. The GRI standards oversight bodies now more clearly parallel the ISO consensus standards process to develop and publish their standards (GRI 2021).

The Sustainability Accounting Standards Board, previously mentioned in Chapter 2, which was absorbed by the International Accounting Standards Board (after having merged with the Integrated Reporting Framework) created seventy-seven sector-specific voluntary standards. These voluntary standards remain voluntary, even as absorbed under ISSB, however not for long. As voluntary standards, in this case are referenced by mandatory accounting standards bodies like the International Financial Reporting Standards Board, they become de facto regulation. Early in 2024, fifteen global jurisdictions announced the adoption of the IFRS ISSB SASB S1 general requirements and S2 climate-related risks voluntary disclosures by 2026 with Brazil mandating the disclosures for their corporate citizens (Corporate Disclosures 2023).

Proprietary Frameworks

There are proprietary frameworks produced by both for-profit and non-profit organizations. Proprietary framework business models may include opportunities for the membership or user-base (organizations) to provide input to the framework, but unlike consensus standards, stakeholder engagement may not be as formal as the ISO process nor shared publicly.

A prominent example of a proprietary framework is Ecovadis' supply chain assessment. Ecovadis invites a participating organization to complete their online assessment questionnaire which is then scored and shared with the organization's supply chain (usually back to their customer). Questionnaires are scored individually and mapped against other organizations in the same sector. Each assessment generates a percentile score and medallion award that can be shared with the organization's customers.

Ecovadis shares their materiality and weighting of certain performance elements, like greenhouse gas emissions, for example, on their website, so users can understand how to interpret results. While individual performance data is not shared, percentile ranking provides an indication of scoring against peers in this proprietary voluntary framework.

Principles

Beyond regulations, consensus, and proprietary frameworks, organizations may also reaffirm their commitment by signing sustainability principles, for example, the United Nations Principles of Responsible Investment, known as the acronym PRI (United Nations Principles of Responsible Investment 2024a).

A signatory to PRI commits to:

- Incorporating ESG issues into investment analyses and decision-making processes,
- Being active owners and incorporating ESG issues into ownership policies and practices,
- Seeking appropriate disclosures on ESG issues by the entities in which they invest,

- Promote acceptance and implementation of the principles within the investment industry,
- Work together to enhance our effectiveness in implementing the principles, and
- Report on our activities and progress towards implementing the principles (UNPRI 2024b).

In addition to signing their commitment, organizations are required to publicly report on their responsible investment activities on an annual basis (UNPRI 2024c).

Initiatives

The final category of general industry frameworks is initiatives. Initiatives are sustainability movements towards a common goal and may have elements of principles and consensus standards. One of the oldest initiatives in sustainability is the Carbon Disclosure Project, now known as CDP due to its expansion beyond carbon disclosures. CDP is a not-for-profit charity celebrating 20 years in 2020. CDP got its start supporting companies to voluntarily disclose their climate impacts in 2000 and has since expanded to deforestation, water security, biodiversity, supply chain partnerships, and plastics for both companies and public entities (CDP 2024). In 2015, CDP launched the Carbon Action Initiative, an investor-led initiative within the CDP reporting framework that demonstrates how organizations within investment portfolios are managing carbon emissions and energy efficiency (Wikipedia 2024). The initiative has three requirements: reduce emissions, disclose reduction targets, and make return-on-investment positive investments in projects. The 2020 CA100+ report showcases the learnings from analyzing over 100 of the signatory investor's portfolio companies and reinforces the need to consider the interconnectedness of sustainability (CDP 2020).

In 2020, CDP launched their 2021–2025 strategic plan, which announced their plans to introduce land, resilience, biodiversity, waste, oceans, freshwater, and food questionnaires to complement the existing climate change, water scarcity, and forests questionnaires (CDP 2022). With these additions, CDP questionnaires will cover all "planetary boundaries and earth systems." CDP teams evaluate all submissions and derive insights that are shared publicly. Collectively, CDP's questionnaires will highlight the disparity between ecosystems demand and the planetary supply. It is the interconnectivity of each of the disclosures that will provide valuable insights, allowing the tracking of our shared progress. The anticipated insights CDP will be generating are anticipated to increase influence with reporting organizations and expand their efforts to reach new reporters. Greater visibility across more sectors is anticipated to also increase public policy and drive action.

ESG Raters

And, last but certainly not least in the vast landscape of frameworks, ESG Raters. ESG Raters are a unique set of proprietary frameworks designed to inform investors as to the ESG risk associated with publicly traded companies – a separate group from ESG indices although in the same sector with potentially the same audience. Analysts working for an ESG Ratings company review publicly available information from organization websites, sustainability reports, and financial reports to create an assessment of the traded company. Similar to other framework types, the number of ESG ratings and rankers has continued to grow. According to ERM's Rate the Raters report in 2020, there were over 600 raters globally. The reliance on such Raters' scoring has led to analysis of the rater's ratings.

To assist organization in understanding the differences between raters, Deloitte publishes an online comparison of the raters (Deloitte 2022). On Deloitte's website, rating agencies are charted to showcase differences in the methodologies in how organization may be scored, subjects covered, and the source of information being evaluated. The website further describes how often an entity's score is updated and the interaction types between the rater and the issuer. Given the variation, it is not surprising the output scoring has also come under scrutiny.

With the increased availability of ESG data and the rise of ESG rating agencies, so has investor criticism of those ESG rating agencies risen. In Harvard Law School Forum on Corporate Governance's Managing ESG Data and Rating Risk, the Forum authors note investors' concerns with "the focus on past performance and lack of predictive value over future performance and the sometimes-divergent opinions of ESG ratings providers for the same company" (Hirai and Brady 2021). Confirming the investors' viewpoint that the assessments are unique to each ratings company, in fact, MIT investigated the divergence of the ESG ratings from six of the most prominent ratings agencies: KLD, Sustainalytics, Moody's ESG, S&P Global, Refinitiv, and MSCI (Reuters 2023). Their work confirmed the divergent results and concluded the raters' overall view of a company influenced the measurement of specific categories (Berg et al. 2019).

The Harvard and MIT work on ESG rater scoring is just one aspect in a broader campaign to align ESG ratings with the EU Commission, also citing conflicts of interest as rating agencies score companies as well as consult on scoring strategies and program implementation with those same companies (EUR-Lex 2024). The EU Commission proposed in June 2023 a regulation on the transparency and integrity of environmental, social, and governance rating activities due to "(i) the lack of transparency on the characteristics of ESG ratings, their methodologies and their data sources and (ii) the lack of clarity on how ESG rating providers operate." Lack of transparency also extends to the ratings that are assessed with only a select few of the ratings agencies posting the ESG scores publicly.

ESG Rankers

We would be remiss if we didn't touch on more visible Rankers in this framework discussion. Rankers may be leading magazines like USA Today or Corporate Knights (as previously discussed in Chapter 2). Organizations may pay particular attention to leaderboards that are most prevalent in their sector. The United Nations also ranks all member states' sustainable development performance, with Finland, Sweden, and Denmark leading the way and South Sudan the final scored member state (noting more than two dozen remained unranked with no numerical score) in the 2023 ranking (Sustainable Development Report 2024). S&P Global ranks the top ESG scoring companies from the over 7500 companies that are assessed using S&P Corporate Sustainability Assessment (CSA) questionnaire and transparently shares the methodology behind the scoring (S&P Global 2022).

Who Is Generally Involved?

Who might be involved in gathering, collating, and disseminating disclosure data will depend on the variety of reporting frameworks your organization may be subject to at any given reporting period. Thankfully, the market does seem to be consolidating.

The frameworks that impact an organization are dependent on a number of factors, including the specifics of the organization and leadership commitments. Determining factors may include, but are not limited to, how the organization is identified:

- Public entity (city or municipality)
- Publicly traded entity (fortune 500)

- Non-profit organization (Red Cross)
- Goods and/or services (performance standards) provider
- Industry or sector within or supporting (mining versus telecommunications)
- Value chain relationship (direct to consumer products (B2C), business-to-business (B2B), suppliers)
- Country or region (EU directives or local regulations)
- Regulation or voluntary standards disclosures applicability (GRI, SASB, CSRD, Ecovadis, CDP)

Local municipalities may have their own policies or ordinances for sustainability performance, often referred to as resilience efforts. The ordinances may apply to local businesses, but local municipalities will also incorporate similar elements into their own internal compliance efforts. For example, a municipality is responsible to ensure safe, accessible drinking water for their residents. The water accessibility must consider climate impacts, such as drought, in the municipal plan to ensure availability during times of water stress. Drinking water quality standards, in the United States, are drafted by the National Sciences Foundation and are consensus standards, but through incorporation into local municipal performance requirements, they become regulatory standards. Due to the complexity of implementing municipal-entity sustainability, often called resilience in the common vernacular of the municipal strategy, a number of roles may be engaged to further the strategy toward disclosures to applicable frameworks. The city manager may oversee the strategic effort. The city engineer will oversee all aspects of infrastructure and drinking water safety. Larger cities may have a dedicated sustainability or health and safety professional to support their programs.

Reporting of sustainability and resilience across municipalities has historically been locality or regionally dependent; however, the public accounting professionals are striving for consistency across disclosures and disclosures for public entities. The International Public Sector Accounting Standards Board (IPSASB) published its first draft public sector sustainability reporting standard for climate impacts at the end of 2024. This aligns with other frameworks and CDP. As CDP has expanded its disclosure effort, public entities have also incorporated it into annual reporting for climate and water security. Just as CDP publishes the list of top-scoring organization in the company category, CDP also publishes the list of top-scoring (A-listed) cities. And, now like private corporations have for over a decade in the EU, public entities will begin disclosures across ESG as well as water security, social responsibility, and resilience.

Publicly traded companies may be subject to frameworks across both voluntary and regulatory, global and industry-specific, and include proprietary frameworks. US-based companies report to the US Securities and Exchange Commission (SEC) on material financial and sustainability issues. The US SEC requires organizations to disclose material human capital issues currently, and once the greenhouse gas emissions and other climate-related impacts litigation is resolved, or the California climate litigation is resolved, disclosure reporting will likely expand.

Whereas in the EU, organizations are required to report on topics across environment, society, and governance for their own entity (CSRD-ESRS) and their supply chain (CDDD). In the United States, organizations follow generally accepted accounting principles (GAAP), whereas many global countries follow the International Financial Reporting Standards (IFRS-ISSB-SASB-S1,S2). The IFRS absorbed the task force for climate-related financial disclosures (TCFD) framework for voluntary climate disclosures and the combined US-based sustainability accounting standards board (SASB). While still voluntary in most countries to report on IFRS sustainability standards S1 and S2, Brazil, and Japan have incorporated mandatory reporting to the standards. Unlike the SEC human capital disclosures, the new sustainability disclosure requirements within IFRS are still voluntary however for most jurisdictions, until 2026.

Corporate organizations are also subject to reporting and disclosure requirements to other frameworks beyond regulatory and financial reporting standards, and in addition to rigorous product performance standards, organizations may be required to disclose voluntary framework disclosures based on their customer-supplier contracts. For example, disclosing to Ecovadis may be voluntary until a customer makes that a requirement of their contract. Some customers may also include the requirement to achieve a certain level of performance to remain in good standing. However, most often, it is the organization's customer that has required disclosure to a specific voluntary framework, for example, CDP or Ecovadis. Due to the customer relationship and sometimes contracts, these voluntary disclosures become a required part of doing business.

Where Is It Interconnected?

Within many of the frameworks, both voluntary and regulatory, there are environmental health and safety disclosures. Depending on the framework and industry sector, disclosures beyond injury rates or the number of environmental spills may be included, for example. The disclosures may be released publicly, that is, as part of a sustainability report or to a regulatory body, but they may also be included in framework disclosures that are shared directly with the organization's customers.

Workplace health and safety metrics, that is, injury rates, may be included as part of a performance requirement in contracts with customers or supply chain partners and are provided through the agreed framework. Additionally, ESG Raters may include occupational injury data as part of their rating system. Industry and sector-specific frameworks, like SASB, may include health and safety metrics and data beyond injury rate, like training information or implementation of an occupational health and safety management system.

Relationships with customers are paramount to successful organizations and the ESG Rater's ratings may influence organizational financial stability and cost of investment dollars (that is, interest rates) (ESG News 2024; Forbes Advisor 2024). Without financial stability and investment, capital expenditures, sometimes necessary to improve safety and health outcomes, may not be as readily available due to increased cost of capital.

Applicability Analysis

In the regulatory arena, an organization would review the regulation to determine if indeed it applied to the organization. When considering frameworks, outside of regulatory reasons, organizations may be struggling with how to determine applicability. As mentioned earlier in the chapter, a voluntary framework may become quasi-mandatory if a customer incorporates the framework by name or its requirements into their contract with your organization. The requirements may become applicable when it makes good business sense to fulfill those requirements – your social license to operate. Sustainability-related operational impacts (energy, waste, and water efficiencies), neighborly goodwill, and community engagement are all seemingly obvious positive aspects of sustainability implementation, but how do you weigh the implementation of SASB over GRI?

The answer may lie within the benchmarking implementation tool discussed later in this book. Benchmarking your organization against others in our sector or industrial value chain can provide insightful information as to what frameworks are resonating with stakeholders in your same value chain.

Similar to US EPA or OSHA regulatory applicability analysis, some frameworks may publish an implementation checklist, for example, early iterations of the EU's ESRS disclosures and metrics were so extensive a spreadsheet was published to assist reporters (EU EFRAG 2024). For organizations using earlier versions of GRI standards, there may be similar spreadsheets for their disclosures still available on the internet as well.

These checklists also ease implementation pressures by allowing the organization to separate implementation into phases based on resources required, degree of difficulty, or other organizational-appropriate measures.

References

Berg, F., Kölbel, J., and Rigobon, R. (2019). Aggregate confusion: the divergence of ESG ratings Forthcoming review of finance (August 15, 2019). https://ssrn.com/abstract=3438533 or http://dx.doi.org/10.2139/ssrn.3438533.

CDP (2020). Analysis of CA100 company data CDP for investor signatories. https://cdn.cdp.net/cdp-production/cms/reports/documents/000/005/312/original/Analysis_of_CA100__Data_for_CDP_Investor_Signatories_v5.pdf?1596046258 (accessed 29 June 2024).

CDP (2022). Accelerating the rate of change CDP strategy 2021–2025. https://www.cdp.net/en/info/about-us/our-five-year-strategy (accessed 29 June 2024).

CDP (2024). https://www.cdp.net/en/info/about-us (accessed 29 June 2024).

Corporate Disclosures (2023). Brazil to mandate ISSB disclosures from 2026. https://www.corporatedisclosures.org/content/top-stories/brazil-to-mandate-issb-standards-from-2026.html (accessed 20 June 2023).

Deloitte (2022). CE table rating ESG English. https://www.deloitte.com/content/dam/Deloitte/ce/Documents/about-deloitte/ce_table_ratings_esg_eng.pdf (accessed 29 June 2024).

ESG News (2024). https://esgnews.com/msci-study-finds-higher-esg-ratings-lower-cost-of-capital-for-companies/ (accessed 11 November 2024).

EU EFRAG (2024). Draft ESRS annex-1 set1 XBRL taxonomy illustrated in excel. https://view.officeapps.live.com/op/view.aspx?src=https%3A%2F%2Fxbrl.efrag.org%2Fdownloads%2FAnnex-1-Draft-ESRS-Set1-XBRL-Taxonomy-illustrated-in-Excel.xlsx&wdOrigin=BROWSELINK (accessed 29 June 2024).

EUR-Lex (2024). Proposal for a Regulation of the European Parliament and of the Council on the transparency and integrity of Environmental, Social and Governance (ESG) rating activities. https://eur-lex.europa.eu/legal-content/EN/TXT/?uri=CELEX%3A52023PC0314 (accessed 29 June 2024).

European Commission Sustainable Finance Package (13 June 2023). https://finance.ec.europa.eu/publications/sustainable-finance-package-2023_en (accessed 20 June 2023).

Forbes Advisor (2024). Customer engagement in 2024 the ultimate guide. https://www.forbes.com/advisor/business/customer-engagement (accessed 29 June 2024).

GRI (2021). The GRI standards a guide for policy makers. https://www.globalreporting.org/media/nmmnwfsm/gri-policymakers-guide.pdf (accessed 1 January 2024).

Harvard Law School on Corporate Governance (2021). Managing ESG data and rating risk. https://corpgov.law.harvard.edu/2021/07/28/managing-esg-data-and-rating-risk/ (accessed 3 April 2023).

International Standards Organization (ISO) (2023). https://www.iso.org/developing-standards.html (accessed 1 January 2023).

Net Zero Initiative (2024). https://www.net-zero-initiative.com/en (accessed 1 January 2023).

Reuters (2023). EU proposes new shake-up of ESG ratings providers. https://www.reuters.com/sustainability/eu-proposes-greater-regulatory-scrutiny-esg-ratings-providers-2023-06-13 (accessed 27 February 2024).

S&P Global (2022). The sustainability yearbook – 2022 rankings. https://www.spglobal.com/esg/csa/yearbook/2022/ranking/ (accessed 29 June 2024).

Sustainability Knowledge Group (2018). What is the purpose of sustainability frameworks and how do they bring value to organizations. https://sustainabilityknowledgegroup.com/what-is-the-purpose-of-sustainability-frameworks-and-how-do-they-bring-value-to-organizations/ (accessed 1 January 2023).

Sustainable Development Report (2024). https://dashboards.sdgindex.org/rankings (accessed 29 June 2024).

United Nations Principles for Responsible Investing (2024a). What are the principles for responsible investing https://www.unpri.org/about-us/what-are-the-principles-for-responsible-investment (accessed 20 June 2023).

United Nations Principles for Responsible Investing (2024b). Public signatory reports. https://www.unpri.org/signatories/reporting-and-assessment/public-signatory-reports (accessed 20 June 2023).

United Nations Principles for Responsible Investing (2024c). Data portal. https://ctp.unpri.org/dataportalv2 (accessed 20 June 2023).

Wikipedia: Carbon Disclosure Project (2024). https://en.wikipedia.org/wiki/Carbon_Disclosure_Project (accessed 29 June 2024).

6

The Capitals

The Capitals

Financial, human, social, natural, and produced capital are referred to as the capitals. In business today, financial capital is the recognized measure of a company's value and is the foundation of capitalism-based economies around the world (Capitals Coalition 2021). In finance, a capital is a "company's assets that have monetary value." This can include inventory and equipment (Investopedia 2024). In this chapter, we will expand on these capitals and how they can form a foundation for new thinking, a "capitals approach," on the connection between human well-being, company value, and economic success (Capitals Coalition 2023). This broader capitals approach can be thought of as the benefits a company receives from people, nature, and its business relationships in addition to a company's traditional financial capital assets that contribute to profit and performance (Nicholles and Seabrook 2023).

Andy Howard, global head of sustainable investment at Schroders suggests, "thinking of capitals as resources, a company's distinctive resources, capabilities and advantage. Distinctive resources could be the quality of a company's talent, the diversity and global breadth of its workforce, the relationships it has with its customers, or how well it has engaged local communities and their trust" (Howard and Seabrook 2023). These distinctive capabilities and resources present an advantage relative to competitors. When a company proactively determines which resources (the capitals) are most important to their business, it can leverage those resources to outperform competitors.

To better understand the relationship of the capitals with society and the economy, let us take a look at their twentieth-century evolution. In post

Connecting the Dots: Environment, Health, Safety, and Sustainability,
First Edition. Kathy A. Seabrook and Tanis J. Marquette.

Second World War, in the western part of the world – especially Europe, building up the economy was also in service to the population. The 1960s, 1970s, and 1980s gave rise to a capitalism where profit maximization came solely to the fore. This was juxtaposed by businesses playing a fundamental role in providing jobs and, in many cases, access to health care through national, state, or private insurance. This provided infrastructure and the flow of goods and services that gave rise to profit (known as financial capital) becoming the focus of a company's value and purpose: to provide a return on investment for the shareholder. The interconnection between financial capital and human, social and natural capital as well as societal value became disconnected from the economy and business in this evolution (Nicholles and Seabrook 2023).

Business stakeholders and societal expectations are beginning to question this disconnect between the capitals, business, and the economy. The capitals approach translates the flow of benefits or the true value a company receives from all available resources and assets to inform decision-making by stakeholders such as investors, insurers, customers, government regulators, employees, etc. Valuation of these capitals can be expressed monetarily or intangibly. In this context, value can be in the form of worker effort and engagement, access to raw materials and natural resources, produced products or services and customer and societal trust and loyalty. "As with their bank balance, a company does not want to deplete these assets (capitals) too quickly. These capitals can be depleted, grown, stay the same or be regenerative," according to Natalie Nicholles, executive director of the Capital Coalition (Nicholles and Seabrook 2023).

Connecting the dots, a company's sustainability, environment, health, and safety initiatives create value for the company and aligns with this capitals approach. Initiatives such as risk assessment, EHS management systems, ergonomics programs, fall protection systems, permitting, spill containment and carbon reduction are designed to mitigate a company's impact and risk to these capitals. These impact risk mitigations preserve human life in the workforce and community, build strong business relationships, and preserve and regenerate the natural environment. Communication is key to influencing business leaders. Table 6.1 highlights the connection between influence, communicating using the language of business, decision-making, and EHS and sustainability professionals.

Definitions of human, natural, social, and produced capital are found in Chapter 16 *Glossary*. In this chapter, the interconnectivity of all the capitals and their contribution to a company's overall value and decision-making are demonstrated.

Table 6.1 Influence and the language of business.

"Human capital is not a term used by the safety profession, but you must speak the language of the person you want to influence. Externally, EHS professionals need to change the conversation. In order to influence people to make the right kind of decisions at the right time." – Malcolm Staves, Global Vice President Health and Safety L'Oréal.

Source: Staves and Seabrook (2023)/with permission of Malcolm Staves.

Social Capital: The Glue That Binds

Good relationships with the local community are an important company asset. Trust, integrity, and reliability are outcomes of building this social capital. As an example, a tech company's local community relationships are a valuable resource to draw upon to attract, train and develop talent for the organization.

According to Natalie Nicholles, executive director of the Capitals Coalition, social capital is sometimes considered "the glue that binds the other capitals together" (Nicholles and Seabrook 2023). In the tech company example above, social capital includes relationships with all stakeholders: customers, suppliers, employees, and workers in addition to the communities in which a company operates. Communities also include customers, the places where employees, operations, and suppliers are located and upon which a company depends. Social capital is also a company's response to as well as its impact on human rights, product quality and safety, consumer data protection and privacy, advertising, product labeling, and other consumer protections.

Another example of social capital is a bank's decision to provide low-cost loans to a socio-economically depressed area in which they operate by providing investment capital for manufacturing and distribution sites that stimulate the economy, create job opportunities, and provide a living wage. As a business, the bank's local community impact is positive, creating value and social capital. The bank could account for this social capital in its profit and loss balance sheet; however, social capital is not often valued as a resource or asset. Using a capitals approach changes the calculus by monetizing, where possible, and incorporating the value of this social capital into a company's profit and loss accounting. When social capital becomes more financially tangible, its influence on decision-making increases. More on value assessment and valuation will be discussed in Chapter 13 *Value: Creation, Assessment, Valuation and Impact Pathways.*

Intangible Becoming Tangible: Paying a Living Wage and the Capitals

Let us take a moment to connect the dots. In Table 6.2 we begin to see the benefits of a company's living wage commitment. in this example, benefits of paying a living wage are created for workers and their families, communities, the company and value chain resiliency as well as society. Benefits include human rights, human well-being, and social and human capital for companies, and their value chain. Paying a living wage is also tied to achieving the United Nations (UN) Global Compact Sustainable Development Goals (UN SDGs).

Social and human capital are critical to business continuity and resilience. This is especially true when severe weather, a natural disaster, pandemic, or an economic crisis causes disruption to a business. During a disruption, the value of social capital is most visible. This is a time when a

Table 6.2 Benefits of paying a living wage.

- While there is no global definition or monetary agreement on a what constitutes a living wage, there is broad consensus on what constitutes a living wage. A living wage is a wage that enables workers and their families to meet their basic needs.
- The construct of a living wage is grounded in the UN SDG 8: *Decent Work and Economic Growth.* SDG 8 advances inclusive, sustainable economic growth through employment, and decent work for all (United Nations Global Compact 2024). The SDGs are foundational to the UN Global Compact, to which many companies around the world are signatories and are committed to paying a living wage to their workers.
- Being paid a living wage enables human rights to food, water, health, housing, education, family life, and fair working hours (SHIFT 2023).
- There is a wider impact on society. Reducing poverty and inequality creates more harmony and social cohesion, which in turn provides a business environment in which a company can be successful (SHIFT 2023).
- Paying a living wage delivers a more mentally and physically healthy, motivated and productive workforce, impacting production output, retention rates, increased resilience, and ultimately financial performance (SHIFT 2023).

Source: SHIFT (2023) and UN Global Compact (2024).

company's relationships with its employees and other workers, suppliers, customers, governments, and investors make a real difference to its resiliency. It is the extent to which customers continue to purchase a company's products, even in the face of disruption and doubt. This customer loyalty and support is the result of built-up trust that the company will make good on its future promises based on past experience. This revenue is a positive, direct impact of the company's social capital.

Let us look at the same disruption scenario described above through a human capital lens: the extent to which an employee or other worker is willing to work beyond the normal requirements of a job during a disruptive event. A specific example is an employee or other worker choosing to work extensive overtime to meet production targets at a manufacturing or construction site, where health-related absenteeism and labor disruption has risen due to the spread of an infectious disease, such as COVID. The decision to work extensive overtime in an emergency situation can be a result of accrued trust in the company and contributes to a company's human capital. Innovation, productivity, efficiencies, and well-being can be the outcome of the employee's or worker's trust in and loyalty to their employer. These examples demonstrate why social capital is the "glue that binds" and why both social and human capital create opportunity, resiliency, and a true measure of a company's value (Nicholles and Seabrook 2023).

The Capitals and a #TrueNorth Purpose

When a company begins to understand how it impacts and depends upon the capitals, it is better able to make decisions more holistically, creating value for the company through nature, its people, society, and the economy (Capitals Coalition 2023). When a company operates with this #TrueNorth purpose/approach and aligns with and drives a capitals approach to holistic decision-making, it passes the #TrueNorthTest (Global Solutions, Inc. and Seabrook 2022a, b, 2024). See Chapter 8 *Beginning the Journey: The Integrated Sustainability Business Model and Sustainability Culture Change Framework* for more information on #TrueNorth and the #TrueNorthTest.

The capitals approach puts impact and dependency risk management at the core of decision-making, which aligns with the intent of managing the environment, social, and economic dimensions of sustainability. When a company integrates a #TrueNorth capitals approach, it takes people, planet and economics into account in valuation, which in turn informs decision-making. Managing impacts and dependencies also aligns with the principles of good risk management, resulting in risk reduction, increased

#TrueNorthTest
(Company Purpose)

'Where the impact and dependency on
People, their Safety, Health & Wellbeing, and the Environment
are valued and reflected in all business decision making
as a measure of
true corporate performance'

- Kathy A. Seabrook, Co-Chair
Capitals Coalition, Valuing Human Capital in OHS Project Group

Figure 6.1 #TrueNorth, a company's purpose. Source: Global Solutions, Inc. and Seabrook (2022a, b) and Global Solutions, Inc. and Seabrook (2024).

opportunities and longer-term safety, environmental, operational, and commercial performance. Figure 6.1 defines #TrueNorth, embodying purpose and a capitals approach.

The Capitals: More Examples of Impacts, Dependencies and Value

Let us further unpack the importance of the capitals and their impact on business and the economy. As we have discussed, if a capital is an asset or resource, it can create or erode a company's wealth or value and be accounted for on the balance sheet (Nicholles and Seabrook 2023). Within companies, this can translate to capital expenditures and budgeting. In an acquisition for example, a privately or publicly held company may need to raise capital by creating debt to finance the acquisition. Accounting for the benefit of and risks to the company's natural, social, and human capital may influence investor decision-making on financing the acquisition.

To better understand the capitals and social capital in particular, the example in Table 6.3 of a mobile (cell) phone company depicts its unrealized social value, which is an element of sustainability.

Sustainability seeks to value externalities; that is a company's impact on people, the environment and economy external to its operations. This can be the social cost of carbon emissions, which is a negative externality to a

Table 6.3 A case for social value creation: a mobile phone company.

Let us use an example of a mobile (cell) phone company to better understand social value.
One can have a variety of apps on their mobile phone. A navigation app can ensure that one never get lost. One globally recognized messaging app currently on the market can act as a "social networking tool" by having 95% of one's friends, family, and colleagues on the messaging app, allowing free connectivity with them. Use of this app also means one can organize to meet up or communicate with one's social network, separately or together, at a moment's notice and for free.
On the personal side, hotels are now using this globally recognized messaging app to provide instant concierge service. This means quick access to restaurant recommendations and instant requests and delivery by the hotel's room service.
One can also do business through this globally recognized messaging app, talking to data or information providers or if a retail provider, sell things through the messaging app. Theoretically, the app can provide an entirely free channel through which to reach customers.
These examples are a positive consequence of the free connectivity through the apps. This is social value to the phone owner and the service provider. However, this social value is not accounted for in the mobile phone service provider's balance sheet.
To further capture this idea of social value, think about how much you would want the mobile phone service provider to pay you to give up these apps or to rescind all access to your phone. Would you expect to give it up for less than the US$50 a month you pay for the mobile phone service? The monetary difference in the monthly cost for service and the value you put on the benefits for free use of the apps is the social value created for a customer and ultimately the mobile phone service provider.
Sustainability seeks to value externalities. Whether it is the social cost of carbon emissions, a negative externality to the company or alternatively, the social value created by free digital connectivity via a mobile phone service provider, it is a positive externality.

Source: Adapted from Bauer and Seabrook (2023).

company. Alternatively, a positive example of an externality is the social value created by free digital connectivity via a mobile phone service provider. According to Angus Bauer, Head of European Equity Research, Schroders, "the social value of connectivity was one of the working groups that I led where we try and actually put a dollar number to the societal value of digital connectivity, the value that goes unpriced." He goes on to say, "one may pay US$ 50 per month for a mobile phone service with unlimited data, domestic phone calls, etc. Does that mean that the total social value associated with the value chain that produces this service is capped at US$50 a month? Or is there a lot more utility that is unpriced and is effectively surplus value created by the mobile phone service provider and the industry?" (Bauer and Seabrook 2023). This positive externality of free digital connectivity can be valued, monetized, and expressed as a percentage of sales to normalize for the size of the company (Bauer and Seabrook 2023).

Bauer puts this in perspective. "At the end of a given year, one can look at its business and surmise which would be true: the world would either give it a bill to pay for the social cost of running the business that is not borne by the balance sheet or profit and loss reports, or provide a check for the social benefit of that does not accrue to you." The difference is the social cost or benefit from the social capital accrued or destroyed. In the case of the mobile phone company, a check would be in order from the social capital gleaned through free digital connectivity. This accounting ultimately provides a truer account of productivity or profitability of a business.

The value of the company and its assets or resources is traditionally determined by an investor or an other stakeholder's valuation process. Using a capitals approach takes into account the impact of risk management and the value created by nature, customers, suppliers, employees, and other worker relationships as well as the ability to produce products and services better than the competition. Proactive accounting for and communicating positive impact and risk mitigation and reduction initiatives can influence stakeholder decision-making in favor of the company.

Another capitals approach example is a beverage manufacturer's dependency on freshwater as a raw material in the production of its products. Access to fresh water, in this case, is the company's natural capital and scarcity of water could mean a significant risk to its financial capital and performance.

An electric vehicle (EV) battery manufacturer is dependent upon natural resources such as lithium, nickel, and cobalt to produce lithium-ion batteries. Access to these metals is a company's asset and natural capital. If these metals are over-mined and become scarce or the use of toxic chemicals in the extraction process reduces employee or value chain worker access to the metals, EV batteries cannot be manufactured, reducing the company's financial capital. Toxic chemicals in the extraction process pose an impact and risk to nature, the health and safety of workers, and financial health of the company. This example demonstrates the intersectionality of all the capitals: human, social, natural and financial.

Because the Capitals approach considers the benefits a company receives from all capitals, these assets can be valued and nurtured. Businesses that think in terms of the capitals as assets, find ways to mitigate the risk of those assets losing value. When the value of access to natural resources, employee productivity and innovation, value chain workers, customer relationships, community trust, and access to capital continue to be lost, eventually, there is no business.

Conclusion and Resources

The Capitals Coalition has published Social and Human Capital and Natural Capital protocols that embrace this capitals approach, focused on decision-making. These protocols guide companies and their stakeholders to better understand the value of all the capitals to their business. They provide decision-making frameworks and valuation assessment processes to guide organizations in accounting for their human, social, natural, and produced capital assets (Capitals Coalition 2019, 2022). By accounting for all the capitals, a business can create opportunities for all stakeholders, in addition to assuring traditional return on investment (ROI) expectations for the shareholders who sit at the core of capitalism. The key to harnessing these opportunities is to integrate this capitals approach and valuation into the business' strategy and decision-making governance, at all levels of an organization. The complimentary Capitals Coalition has also published an online Coursera course providing an in-depth review of these decision-making frameworks. See the Capitals Coalition Coursera link in the reference at the end of this chapter (Capitals Coalition 2023).

In addition, Chapter 13 *Value: Creation, Assessment, Valuation and Impact Pathways* provides more information on value and valuation assessment.

References

Bauer, A. and Seabrook, K.A. (2023). October Interview with Angus Bauer, Head of European Equity Research, Schroders. Bauer, Angus (interviewee) and Seabrook, Kathy A. (interviewer).

Capitals Coalition (2019). Social and human capital protocol. https://capitalscoalition.org/capitals-approach/social-human-capital-protocol/ (accessed 8 August 2023).

Capitals Coalition (2021). Social and human capital protocol primer. https://capitalscoalition.org/wp-content/uploads/2021/12/Social-Human-Capital-Protocol-A-Primer-for-Business.pdf (accessed 17 August 2023).

Capitals Coalition (2022). Valuing human capital in occupational health & safety. https://capitalscoalition.org/publication/valuing-human-capital-in-occupational-health-safety/ (accessed 17 August 2023).

Capitals Coalition (2023). Valuing nature and people to inform decision-making. Coursera Online Course (complimentary). https://capitalscoalition.org/capitals-approach/training-resources-old/ (accessed 18 December 2023).

Global Solutions, Inc. and Seabrook, K.A. (2022a). Connecting the Dots: Sustainability, ESG, Human Capital and the Safety Professional presented at the Board of Certified Safety Professionals Global Learning Summit. Virtual (11 May 2022).

Global Solutions, Inc. and Seabrook, K.A. (2022b). The EHS Professionals Expanding Role in Human Capital, Sustainability & ESG (8 December 2022). American Society of Safety Professionals Webinar.

Global Solutions, Inc. and Seabrook, K.A. (2024). Connecting the dots: the value of people at work and their health, safety and well-being. *3 May 2024 Keynote Presentation at the 34th International Congress on Occupational Health*. Marrakesh, Morocco.

Howard, A. and Seabrook, K.A. (2023). 19 October 2023 Interview with Andy Howard, Global Head of Sustainable Investment, Schroders. Andy Howard, Andy, (interviewee) and Seabrook, Kathy A. (interviewer).

Investopedia (2024). Capital: definition, how it's used, structure, and types in business. https://www.investopedia.com/terms/c/capital.asp#:~:text=Capital%20is%20typically%20cash%20or,Open%20a%20New%20Bank%20Account (accessed 12 June 2024).

Nicholles, N. and Seabrook, K.A. (2023). 22 September 2023 Interview with Natalie Nicholles, Senior Director, Capitals Coalition. Nicholles, N. (interviewee) and Seabrook, Kathy A. (interviewer).

SHIFT (2023). Accounting for a living wage. https://shiftproject.org/accounting-for-a-living-wage/#:~:text=What%20is%20a%20living%20wage,family%2C%20including%20some%20discretionary%20income (accessed 10 October 2023).

Staves, M. and Seabrook, K.A. (18 May 2023). Interview with Malcolm Staves, FIChemE, Global Vice President Health and Safety, L'Oréal (interviewee) and Seabrook Kathy A. (interviewer).

United Nations Global Compact (2024). Living wage. Taking the living wage tool. https://unglobalcompact.org/what-is-gc/our-work/livingwages (accessed 14 June 2024).

7

Emerging Topics

Introduction

After a career spanning decades, the authors have experienced issues emerge. Emerging issues, in hindsight, are material issues that have been seemingly bubbling underneath the radar, talked about within smaller professional groups, until they become more material and are elevated to critical mass. Or put another way, many topics have been known for some time and the growing interconnectivity to adjacent sectors or issues creates the opportunity for the issues to become more widely understood and then rise to the material level across enough sectors to "emerge." At the forefront of many people's minds today is artificial intelligence; however, Our World in Data lists this technology's first use in the 1940s.

To understand this emerging issue and its immediate challenges for health, safety, and sustainability, the authors interviewed John C. Havens, esteemed author, former actor, and thought leader on social and emerging issues. He explains that artificial intelligence (A.I.) uses various methodologies utilizing combinations of human (aka "personal") and planetary data. He goes on to note that the use of A.I. without a fundamental understanding of ethics is like working in any system or profession (like health, safety, and sustainability) without a clear definition or parameters of what "good" means. For instance, where GenAI models (or methodologies using "Large Language Models" like ChatGPT) use massive amounts of water and energy where it is unclear how much of these resources are being used, the designers or users of AI can't clearly state the depth of harms to the planet and people. This is where designers, businesses, and policymakers need to ask, 'how can we ensure that what we're creating will accountably improve

Connecting the Dots: Environment, Health, Safety, and Sustainability,
First Edition. Kathy A. Seabrook and Tanis J. Marquette.

human and planetary flourishing by using metrics like SDGs or ESGs at the outset of design?

Plus, it's critical to ask if AI designers are aware of cultures and languages beyond their own. For instance, many AI systems come from Western sources which also feature Western ideals on law as well as ethics. In Japan, there's an ethical tradition known as Shinto. In Africa, there's an ethical tradition known as Ubuntu. "Not recognizing these traditions means AI systems will more than likely cause harm where this can be avoided by factoring in the nuances of these cultures and overall communication at the outset of design" (Havens and Marquette 2024).

As John shares, to truly understand the holistic value of AI, we must not only understand how data is used in these systems but recognize and be accountable toward the underlying basis and biases embedded within the cultures creating the models. We must instill models for applied ethics through the creation of foundational policies and practices to ensure A.I. systems are designed for the accountable benefit of all people and our planet.

> Just transition is a framework developed by the trade union movement to encompass a range of social interventions needed to secure workers' rights and livelihoods when economies are shifting to sustainable production, primarily combating climate change and protecting biodiversity. – Adapted from United Nations Development Programme (2022)

As corporates have become more engaged with stakeholders around climate, readers may have also heard the term "just transition" another intersectionality or interconnected sustainability adjacency that became elevated both for organizations transitioning energy mix and both employees working for those companies and the communities in which those organizations were located. As the organizations re-evaluated and re-adjusted their portfolio, business decisions impacted their local workforce and community. The authors first heard this term from an oil & gas colleague, but since it too has emerged to other sectors (Climate Justice Alliance 2024).

> The United States EPA States that "Environmental justice" means the just treatment and meaningful involvement of all people, regardless of income, race, color, national origin, Tribal affiliation, or disability, in agency decision-making and other Federal activities that affect human health and the environment so that people:
>
> - are fully protected from disproportionate and adverse human health and environmental effects (including risks) and hazards, including

those related to climate change, the cumulative impacts of environmental and other burdens, and the legacy of racism or other structural or systemic barriers; and
- have equitable access to a healthy, sustainable, and resilient environment in which to live, play, work, learn, grow, worship, and engage in cultural and subsistence practices.

Adapted from United States Environmental Protection Agency (2024), https://www.epa.gov/environmentaljustice, last accessed 29 July 2024.

The just transition and its focus on societal impacts of energy transition opens an eye to another adjacent impact environmental justice. Again, environmental justice (EJ) is not at all new; however, it is still not mainstreamed into all permitting requirements and from that sense, may feel like it is emerging. The US EPA's dedicated EJ website supports grant funding, permitting navigation, history, and various resources to promote the elevation impacted peoples toward removal of historical barriers burdening healthy, resilient communities.

Communities are also becoming more resilient by focusing on reintroducing nature. Whether through biodiversity, nature-based solutions (NbS), or rewilding, three related by different terms, nature is emerging in our communities – even our city-communities. Biodiversity (the 2022 focus of COP15) the 15th United Nations conference of the parties led to the agreement to protect 30% of land and oceans by 2030 in Montreal, Canada. To add context, the United Nations released the Our Common Future Report, known more commonly as the Brundtland report in 1987, but it was at COP15 that it was acknowledged that biodiversity and nature are a critical component to meeting the goal of limiting global warming to 1.5 °C. The International Union for Conservation and Nature (IUCN) defines a NbS as one that addresses societal challenges through actions to protect, sustainably manage, and restore natural and modified ecosystems, benefiting people and nature at the same time. As organizations are implementing their sustainability strategies, you may see initiatives that incorporate NbSs into projects and programs for the benefit of the local community and organization. And the final nature-related emerging topic the authors have seen is rewilding. This emerging concept is occurring both in residential neighborhoods and commercial properties (as a measure to conserve water and bring back pollinators), but also across municipalities and that is rewilding. Rewilding is the practice of allowing, or purposely causing a landscape space, to revert back to nature. For example, corporate parking lot may revert back to wildflower grass space or a mown grass space may be allowed to grow freely. This rewilding decreases the amount of watering required, returns the land to its natural vegetation, and increases the potential for natural pollinators to return (IUCN 2024).

Topics are constantly evolving as to what is front and center across the discussion . . . however, the old adage, the more things change, the more they stay the same applies. Most emerging topics will be old news somewhere in the sphere of sustainability and over time different topics ebb and flow. The new topics to pay particular attention to happen when issues, topics, solutions, and technology collide. Just as artificial intelligence has been around since the mid-1900s, and sustainability has been around nearly the same time when the two are entwined . . . . This is a new and novel approach to solving the solutions of tomorrow.

Who Is Generally Involved?

Environmental, health, safety, sustainability and risk professionals have been at the forefront of people and technology as they lead culture change for their organizations. As practitioners and professionals innovate to reach workers on new topics in new ways and bring solutions to the work environment, technology is there to assist. One such solution is "wearables" bringing data and information back to the health and safety professional to support workplace safety improvements. Wearables do not begin to describe the breadth of new technology increasing information available to the professional. The list includes smart hard hats, exoskeletons, worker ergonomics, lone worker monitoring, augmented glasses, environmental monitoring, proximity monitoring, and worker vital sign monitoring (Slate Safety 2024). In addition, as we solve for worker health and safety, A.I. is working to solve for a wide ecosystem of health and safety . . . Innovation, emerging material issues, and solutions coming together.

In Europe, BeeODiversity is using biomonitoring and bioindicators to measure pollution through the analysis of bee pollen collected from the bees acting as natural drones. This novel technique of combining nature and technology allows scientists to understand industrial and agricultural pollution and biodiversity over a significant range of land, reducing the health and safety risk to humans that would have otherwise had to collect environmental samples across mountainous landscapes. The award-winning capability and exciting new technology were an outgrowth of the founder's investigation into the disappearance of bees (BeeODiversity 2024).

Where Is It Interconnected?

As organizations strive to solve for tomorrow's challenges today, environmental health safety and sustainability professionals are instrumental in charting the pathways to future success. Built on the foundations of sound risk management and systems safety programs and the hierarchy of controls, health and safety professionals routinely evaluate new products and services, conduct process hazard evaluations, and process safety reviews as part of their roles (Board of Certified Safety Professionals 2024). As new products and technologies of the future emerge and new roles and professions are created from them, professionals and practitioners will be instrumental in supporting compliance, hazard assessments, and permitting processes to ensure their organizations and people remain safe and competitive in this changing environment.

As the front line to anticipate hazards, managing risk, and establishing appropriate controls and programs to effectively protect workers and the environment, the role of environment, health and safety, and sustainability professionals and practitioners could not be more critical in the discussion of emerging contaminants, technologies, and the solutions of the future (ASSP 2023; AIHA 2024).

Why It Has Value?

Effectively managing environmental and human protection while achieving greater heights with stakeholders, and competitive advantage if the organization is a company or community, is the pinnacle of success. Staying ahead of emerging topics, technology, and products and services, ensures a greater opportunity for your organization to effectively manage risks, but also to capitalize on opportunities (insight & foresight 2024).

References

American Industrial Hygiene Association (2024). Protecting people https://www.aiha.org/ih-careers/discover-industrial-hygiene (accessed 4 July 2024)

American Society of Safety (2023). Safety Professionals Handbook 3 Haight, Joel Editor American Society of Safety.

BeeODiversity (2024). Preserving and regenerating biodiversity https://beeodiversity.com/en/who-we-are/our-history/ (accessed 30 June 2024).

Board of Certified Safety Professionals (2024). Certified safety professional https://www.bcsp.org/certified-safety-professional-csp (accessed 30 June 2024).

Climate Justice Alliance (2024). The just transition https://climatejustice-alliance.org/just-transition/ (accessed 24 June 2024).

Havens, J.C. and Marquette, T.J. (31 January 2024). Interview with John C Havens, Executive Director of The IEEE Global Initiative for Ethical Considerations in Artificial Intelligence and Autonomous Systems and The Council on Extended Intelligence. Havens, John C (interviewee) and Marquette Tanis J (interviewer).

insight & foresight (2024). https://www.insightandforesight.com.au/blog-foresights/how-to-stay-ahead-rapidly-changing-world (accessed 11 November 2024).

IUCN (2024). Benefits and risks of rewilding https://www.iucn.org/resources/issues-brief/benefits-and-risks-rewilding (accessed 21 June 2024).

Our World Data (2022). The brief history of artificial intelligence: the world has changed fast — what might be next? https://ourworldindata.org/brief-history-of-ai (accessed 30 June 2024).

Slate Safety (2024). Top safety wearable technology for 2024 https://slatesafety.com/top-safety-wearable-tech-2024 (accessed 30 June 2024).

United Nations Development Programme (2022). What is just transition? And why is it important? https://climatepromise.undp.org/news-and-stories/what-just-transition-and-why-it-important (accessed 30 June 2024).

United States Environmental Protection Agency (2024). Environmental justice https://www.epa.gov/environmentaljustice (accessed 30 June 2024).

Section II

Implementation Tools and Resources

8

Beginning the Sustainability Journey

Integrated Sustainability Business Model

In this chapter, we introduce the *Integrated Sustainability Business Model* (*Model*), which provides implementation tools to guide companies through sustainability strategy development, action planning, and implementation – no matter what stage they are in their sustainability journey. The *Model* is depicted in Figure 8.1 and incorporates the *Sustainability Culture Change Framework (Framework)*, the *Sustainability Strategy and Decision-Making Roadmap (Roadmap)*, and various sustainability assessment tools. At the core of the *Model* is integration and implementation guidance on managing and reporting on sustainability-related, human capital and EHS risks, impacts and opportunities.

Collectively and organizationally, the *Sustainability-Related Business Model* focuses a company on its current and ideal future state of risks and opportunities as well as impact management and sustainability performance. It highlights opportunities for integration, adaptation, efficiencies, value creation, and resilience.

In the *Model's* implementation hierarchy, the first implementation tool is the *Sustainability Culture Change Framework ('Framework')*. The *Framework* incorporates a holistic approach to strategy and execution, building upon basic principles of organizational development, change management, and business integration. Integrating sustainability into the business can be an adaptive organizational change that will be further discussed in this chapter.

The *Sustainability Strategy and Decision Making Roadmap ('Roadmap')* is the second implementation tool in the hierarchy. It provides a detailed, sequential process to develop and execute strategy and action planning. The

Connecting the Dots: Environment, Health, Safety, and Sustainability,
First Edition. Kathy A. Seabrook and Tanis J. Marquette.

Figure 8.1 The integrated sustainability business model. Source: Global Solutions, Inc. and Seabrook (2024a, 2024b).

Roadmap and additional assessment and implementation tools will be discussed in more depth in Chapter 9 *The Sustainability Strategy and Decision Making Roadmap.*

Let us break down the *Model*, beginning with a discussion on the *Framework* highlighted in Figure 8.2.

Sustainability Culture Change Framework (Framework)

The implementation sequencing of the *Integrated Sustainability Business Model* (Figure 8.1) begins with the *Sustainability Culture Change Framework ('Framework')* highlighted in Figure 8.2. The full *Framework* is depicted Figure 8.3 and is grounded in recognized organizational development and change principles. It can be used by companies who seek to successfully navigate an ambiguous and disruptive business environment that requires agile, engaged and empowered leaders and workforce. Aligning and integrating sustainability into the DNA of a company by engaging and empowering its decision makers, assures sustainability is considered in decision-making at all levels of the organization. When this occurs, there is a greater likelihood that sustainability initiatives are leadership led and integrated into the company's purpose, culture, values, strategy, and governance structures.

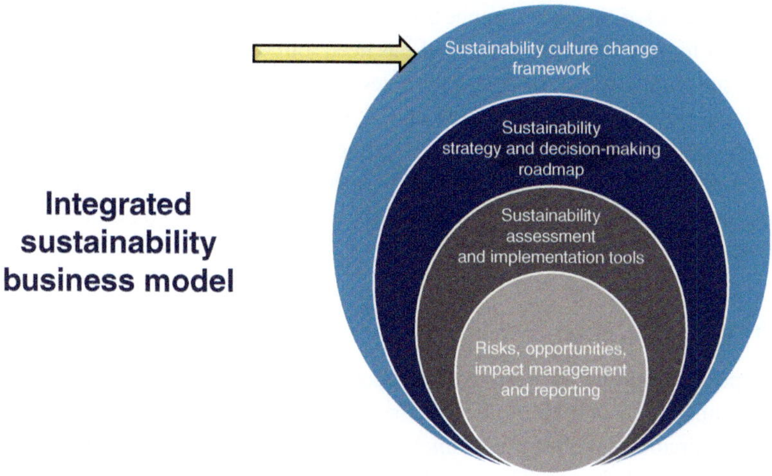

Figure 8.2 The integrated sustainability business model: the sustainability culture change framework. Source: Global Solutions, Inc. and Seabrook (2024a, 2024b).

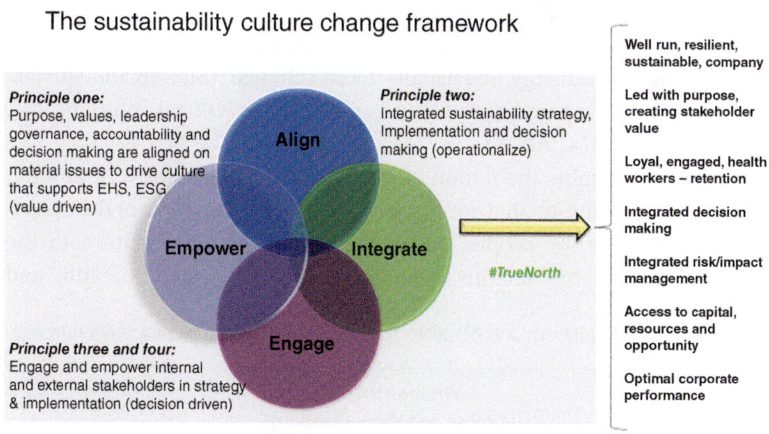

Figure 8.3 The sustainability culture change framework. Source: Global Solutions, Inc. and Seabrook (2024a, 2024b).

The foundation for this Sustainability Culture Change Framework began with this coauthor's collaborative work in the steel industry, applying recognized principles of organizational leadership and development, coupled with implementation experience using EHS integration tools and techniques. This co-author, Kathy Seabrook, wishes to recognize and acknowledge Tom

Reeves, CEO of Engaged Change Solutions and Mike Oberhill, principal consultant for our collective work that has been foundational to the creation of this sustainability culture-focused implementation Framework.

Exploring the four principles of the Framework will better help you understand what scalable, effectively integrated sustainability looks like in a company. The following breaks down the *Framework*, its principle elements and applicability toward advancing a company's sustainability initiatives and overall journey.

Principle One: Align

Purpose, #TrueNorth, #TrueNorthTest, Good Governance and Change

A company's sustainability journey begins with recognizing the integration of sustainability within the business and its current processes is a complex organizational change. The extent of change is dependent upon where the company is in its sustainability maturity. Therefore, navigating this type of change requires a company to identify and align on a common understanding of its purpose and the current state of sustainability integration. According to Dave Stangis and Katherine Valvoda Smith in their book, *The Executive's Guide to 21st Century Corporate Citizenship*, "Purpose is the motivation for all it (the company) does...purpose is the reason you're undertaking your strategy, and usually doesn't change" (Stangis and Valvoda Smith 2017). A company's purpose is its #TrueNorth (Global Solutions, Inc. and Seabrook 2022a, 2022b).

Table 8.1 highlights the notion of #TrueNorth in the context of an integrated sustainability-related approach. #TrueNorth incorporates the contribution, the value, of people, business relationships, and nature to the company and its brand. This includes the value of safety, health, and

Table 8.1 The #TrueNorth and #TrueNorthTest, integrated into a company's purpose.

#TrueNorthTest **(Purpose and Value Creation)** "Where the impact and dependency on People and their Safety, Health & Wellbeing, and the Environment, are valued and reflected in all business decision making as a measure of true corporate performance." Kathy A. Seabrook, Co-Chair Capitals Coalition, Valuing Human Capital in OHS Project

Source: Global Solutions, Inc. and Seabrook (2021, 2022a, 2022b, 2022c, 2024a, 2024b, 2024c).

well-being to people working in a company and its value chain when measuring corporate performance (Global Solutions, Inc. and Seabrook 2021, 2022a 2022b, 2022c, 2024a, 2024b, 2024c). By definition, this premise of #TrueNorth aligns with a company's purpose and objective of longer-term value creation for all stakeholders, not just shareholders, which has been the traditional purpose of a company (Capitals Coalition 2021; US Business Roundtable 2019).

This co-author's notion of #TrueNorth purpose emanated from the US business roundtable's 2019 statement on the purpose of a corporation. In this published statement, 180 CEOs from major US companies signed a statement of purpose advocating a company should create value for all stakeholders if they are to succeed over the long term (US Business Roundtable 2019). This redefined purpose meant companies should "serve their shareholders, deliver value to their customers (including general consumers), deal fairly with their suppliers, invest in their employees and support the communities in which they operate" (US Business Roundtable 2019). This was a game changer for aligning sustainability with a company's purpose, its #TrueNorth (Global Solutions Inc. and Seab rook 2021, 2022a 2022b, 2022c, 2024a).

#TrueNorth directionally aligns and integrates good governance, transparency, and accountability systems to enhance alignment and continuity of a company's purpose and decision-making. This enhances its longer-term value, profitability, and success. Well-functioning governance structure enables consistent decision-making in the face of risks, opportunities, stakeholder expectations, ethical dilemmas, crisis scenarios, and compliance obligations. Governance structure incorporates company policies, processes, procedures, 'institutional memory', and organizational clarity on responsibility and accountability systems. See Table 8.2 for examples of company governance structure policies, processes, and procedures incorporating responsibility and accountability systems.

Table 8.2 Examples of governance structure policies, processes, and procedures.

Governance Structure Policies, Processes and Procedures
• Criteria for fiduciary and other decision-making
• Crisis management policies and procedures
• Code of conduct
• Conflict of interest Disclosure Policy (Director and officer)
• Criteria for board structure and composition
• Internal and external audit
• Equitable talent development and succession planning
• Whistleblower protection policies and practices

The International Standards Organization's management systems standards incorporate the high-level structure, which provides a governance structure for managing material risks such as quality of medical products, hazardous materials and waste management, information security and data protection, health and safety of workers, asset management, etc. We will further connect the dots on management systems integration in Chapter 9 *The Sustainability Strategy and Decision Making Roadmap,* where the *Integrated Impacts, Risks and Opportunities Management Cycle* is introduced.

Recognizing sustainability integration as an organizational change, and building ambiguity and agility capacity in the face of change, will increase the organizational ability to pivot and adapt in a timely way. The catalyst for change in the sustainability space can present in varying ways: market demand, customer expectations and requirements, regulatory requirements, and internal leadership decisions. A leadership decision by the CEO of an additional 20% target reduction in water usage at all manufacturing sites in the next three years is a sustainability-related catalyst for change (in natural capital).

The capacity and ability to pivot and adapt in the face of change is influenced by how well a company has aligned and integrated sustainability into its business policies, processes, practices, and procedures. It is also the extent to which its people are engaged and empowered to make decisions. These notions are at the core of the *Framework.*

Individual and organizational agility embraces innovative responses to unpredictable change. This is especially true as the sustainability space continues to evolve in the areas of mandatory and voluntary standards, labor and supply chain disruption, recognized dependency on natural resources, and customer and investor expectations.

With a better understanding of this intersectionality of purpose, #TrueNorth, good governance, and change, let us explore ways to imbed this in the culture and ethos of a company.

Influencing Begins with a Compelling "Why"

What's Your Personal "Why?"

Before embarking on a sustainability journey with your company, it is important to determine your personal "Why." Why is championing sustainability for your company important to you? By answering the "what's your personal why" question for sustainability and EHS, you set up a foundational "why" for every conversation you will have with your company's internal and external stakeholders. This is key to influencing decision-making on sustainability initiatives. This applies to conversations with front-line workers manufacturing the company's products or providing services, operational leaders as well as the c-suite, board of directors, customers, suppliers, regulators, rating agencies, and investors. When one truly believes an initiative is important and valued, what follows is a next-level commitment to personally engage, actively champion, and support the initiative. This is particularly true in the face of doubt and challenge . . . and organizationally, you will experience plenty of this type of resistance along the way! The more you understand and can articulate your personal "why" in your sustainability conversations, the more they become values driven and compelling. This is a powerful influencer (Global Solutions, Inc. and Seabrook 2022c, 2024a).

What Is the Company's "Why?"

Now that you have established "Why" sustainability is important to you, the next step is to understand the company's "Why." Keep in mind the ultimate goal of implementation is to imbed sustainability into decision-making throughout the company, its purpose, values, strategy, culture, policies, processes, practices, procedures, etc. To do this requires aligning sustainability with the business elements identified in Table 8.3.

For business leaders, sustainability must create value for the company and its stakeholders; this includes shareholders for publicly traded companies. Also note privately held companies seeking capital for acquisitions or other capital expenditures may also be asked to meet investors' sustainability (ESG) expectations. Value creation can be a result of customer and worker trust and loyalty, brand integrity, market share, revenue generation from new products and services and access to natural resources and capital. See Chapter 3 *Sustainability and ESG Key Concepts* for more information on value, value creation, and valuation.

Table 8.3 Aligning sustainability.

Aligning Sustainability and Business Elements
• Purpose • Values • Leadership expectations • Strategy • Culture • Governance and accountability systems • Material risk, opportunity, and impact management (includes impacts on people, business relationships, and the environment)

A company's ability to embrace and communicate why sustainability is important is where the company's #TrueNorth - sustainability journey begins. According to Marty Linsky and Ronald Heifetz, Harvard University Kennedy School of Government professors and coauthors of *Leadership on the Line: Staying Alive through the Dangers of Change,* in order to lead through change, "leadership cannot be disconnected from purpose" (Heifetz and Linsky 2002). When company leaders commit and champion a compelling "why" sustainability is important, that commitment drives purpose, culture and influences decisions throughout the organization. It also inspires followers by providing a sense of purpose. Telling a captivating story about why sustainability is important motivates followers to action, especially when it is personal to them (Seabrook and Reeves 2018). This is a leverage opportunity for the sustainability and EHS professional, and a solid foundation upon which to begin a company's sustainability journey.

"The role of the EHS professional is evolving and is requiring a skill set beyond engineering. Understanding systems thinking, plan-do-check-act and how to integrate an initiative into the business, and deliver an implementation plan are imperative skills," according to Malcolm Staves, Global VP Health and Safety, L'Oréal. This particularly holds true for new initiatives and valuation approaches to sustainability and the capitals (see Chapter 6 *The Capitals*).

When interviewing your business leaders to better understand their reasons for incorporating sustainability into the business, consider the questions in Table 8.4. This data-gathering exercise collectively focuses the company and its leaders on the importance of sustainability to the company's purpose, strategy, and performance. This in turn influences corporate culture and resulting decision-making that supports current and future sustainability initiatives and opportunities. Leader's answers to these guiding questions provide business intelligence on the company's current state of sustainability. The output of these interviews and discussions will help to leverage new and existing company sustainability-related initiatives. These initiatives will positively impact future business opportunities, inform sustainability strategy and action planning, and enhances overall business performance if they are integrated well.

The Decision Makers

Referring back to the *Framework* in Figure 8.3, principle one aligns sustainability with the requirements and expectations of internal and external decision makers and stakeholders. This principle continues to help a company identify its current and ideal future state of sustainability. Understanding a company's current state and motivation for beginning a

Table 8.4 Getting to "why:" company sustainability leadership questions to consider.

Getting to "Why:" Sustainability-related leadership questions
1 Is the company publicly traded or privately held?
2 Small- and medium-size companies, global multinational
3 What industry sector?
4 Number of locations, area, and location of operations?
5 Centralized or decentralized leadership, management, and decision-making?
6 What is the publicly communicated purpose of the company? What is its mission and values? How are they communicated internally and externally?
7 Why is sustainability a company imperative?
8 Who are the company's stakeholders? What is important to them? Their requirements and expectations?
9 What does the supply and value chain look like? Sole sourced products or services? Geographic locations? Supply chain disruptions, labor shortages?
10 What is important to the company?
11 Beyond Environmental. Are risks and impacts to people, customers, suppliers, communities, and the environment identified and managed? (social, human, natural, and produced capital)
12 Worker safety, health, and well-being assessment included in sustainability-related materiality?
13 Are sustainability- related risks and impacts material and considered in business decision-making?
14 Are material risks and impacts reported? Publicly, privately?
15 What are the company's compliance obligations? (i.e. ESG, sustainability, Securities and Exchange Commission (SEC) or other financial securities authorities, Department/Ministry of Labor, discrimination laws (i.e. US Equal Employment Opportunity Commission), occupational safety, health, environmental regulations, product safety, data protection, Food and Drug Administration, Consumer Products Safety Commission, European Union and other country regulations and directives, supply chain accountability, human rights, department of transportation, US Homeland Security, etc.)
16 What do the competitors look like? Who are they, what is their position in the market, value chain, and customer base? What is their material risk profile? How do they compare relative to sustainability risk and impact management? Are there similar customer requirements and expectations? Does your company have a competitive advantage?
17 What are the business drivers for identifying, assessing, measuring and managing sustainability-related material impacts?
18 Are there mandatory requirements for public disclosure? Benefits for public disclosure that are not mandatory?
19 Ultimately, "why" should the company begin a sustainability journey?

sustainability journey requires continually engaging business leaders and company stakeholders to proactively monitor changes in their sustainability expectations and requirements. This enables continual alignment on expectations, requirements, purpose, and #TrueNorth, influencing culture, values, strategy, priorities, implementation objectives and targets, KPIs, and timely communication on measurable progress. This supports sustainability initiatives.

When asked what advice he would give to sustainability professionals, Michael Cooke, a seasoned global sustainability professional, counsels, "don't assume the business leaders are fully aligned with your sustainability strategy or that they understand how sustainability initiatives contribute to the business." He goes on to say, "they (business leaders) will be getting pressure from many customers and questions from investors, so it is important that you show how you can help them "look good" through explaining how your strategy links up with the questions they are being asked." Ultimately, he continues, "you will then be invited to have regular discussions with your business leaders and give them the needed feedback to help with those customer or investor driven sustainability-related discussions on an ongoing basis."

Time spent with decision makers, and those who influence them, is essential to gathering a broad perspective on why sustainability is or is not important to them, the company and its stakeholders. An initial stakeholder mapping exercise is an essential tool to determine the key stakeholders with whom to engage and assess the current state and an ideal future state of sustainability in the company.

Table 8.5 provides examples of Internal and external stakeholder decision makers. Note: internal and external stakeholders can vary based on the company, its size and organizational structure, nature of operations, number of sites geographic location, industry sector, risk profile, value chain, and impact and dependency risk factors.

Chapter 10 *Stakeholders and Stakeholder Mapping* provides more information on the stakeholder mapping process noted above.

Aligning Purpose, Strategy, Integration, and Performance Excellence

When a company's purpose, culture, strategy, policies, process, and procedures are directionally aligned with internal and external sustainability stakeholder requirements and expectations, the value and positive impact a company has on people, the environment and natural resources emerge through sustainability initiatives (inputs and outputs). Chapter 4 *Theory of Change* provides additional information on the correlation between inputs, outputs, outcomes, and impact. Chapter 3 *Sustainability and ESG Key*

Table 8.5. Examples of internal and external stakeholder decision makers.

Stakeholder decision makers	
Internal decision makers	**External decision makers**
• The board of directors • Chief executive officer • Finance • C-suite • Other company leaders at all levels (i.e. operations, workers and their representatives, legal counsel, investor relations, corporate communications, EHS, quality, research and development, human resources, procurement, logistics maintenance, etc.)	• Customers • Suppliers • Investors • Insurers • Rating agencies • Applicable nongovernmental organizations (NGOs) • Government regulators (i.e. securities and exchange commission, food and drug, safety, environmental, labor, hazardous materials, national security entities, etc.) • Other

Concepts continues to make the connection between impact, value, and valuation using the impact pathways process and delving into valuation examples. Impact pathways do this by incorporating inputs, outputs, and outcomes, resulting in impact and value creation (monetized and non-monetized) for various stakeholders.

As noted in Figure 8.4, purpose drives culture, strategy, and an integrated approach to implementation focused on reducing risk and increasing positive impacts. This chapter began with the premise that a company's purpose is its #TrueNorth. This purpose enables opportunities, positive impact, and reduction of risk to people, the environment, society, communities, customers, the value chain, and other business relationships and stakeholders, including the business itself. When a company's purpose is clearly defined, aligned, communicated, and integrated into its strategy, it also influences integrated decision making, which is foundational to a company's culture and commercial, operational, sustainability, and EHS excellence (Seabrook 2023). When company policies, systems, processes, and procedures are also aligned and integrated into the business and operations, they enable longer-term impact and dependency risk management as well as support its #TrueNorth purpose. Table 8.6 further illustrates these relationships and interconnectivity (Brandon and Seabrook 2023).

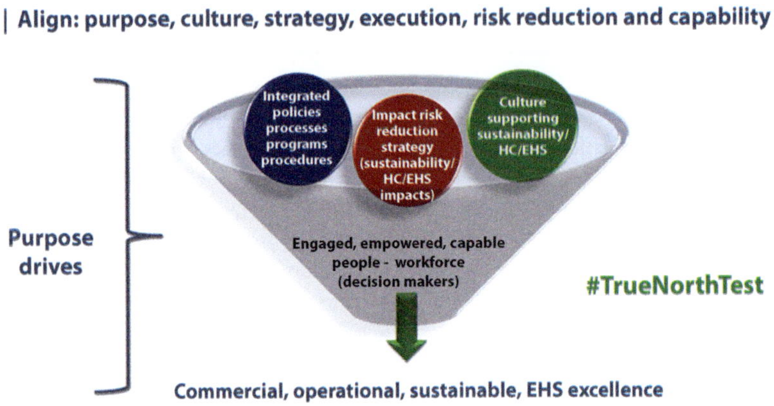

Figure 8.4 Aligning purpose, culture, strategy, execution, risk reduction and capability. Source: Global Solutions, Inc. and Seabrook (2022c, 2024a, 2024b, 2024c).

Table 8.6 A sustainable company: sustainability strategy, program, policies, processes and procedures.

A sustainable company
The sustainability program or process is the actual effort or work an organization does to be sustainable. The actual effort that's occurring may be the development of a strategy, policies and procedures to decarbonize or executing on HR policies that enhance the company's human capital while benefiting the "humans that are the capital."
– Chet Brandon, Director, Health, Safety, & Environmental Protection GrafTech International Holdings Inc.

Source: Brandon and Seabrook (2023)/with permission of Chet Brandon.

It is important to distinguish between what it means to be a sustainable company and the company's actual sustainability policies, processes, programs, and procedures. Table 8.6 makes a clear distinction between the two, while also demonstrating their interconnectivity. Organizational understanding of these concepts is integral to aligning the company and its leaders on purpose, culture, strategy, risk reduction, the four *Framework* principles, and capability. This facilitates sustainability efforts throughout the company in order to reap the benefits of commercial, operational, sustainability, and EHS excellence.

Principle Two: Integrate

Generally speaking, integration means to bring together different elements to create a cohesive whole. In the context of sustainability, integration brings together elements of a company's purpose, culture, strategy, action planning, governance structure, decision-making processes and internal stakeholders (including workers) in an effort to seamlessly optimize business and sustainability performance. Integrating sustainability into the business and its processes creates value for the company and its stakeholders. We will unpack this notion in Chapter 13 *Value: Creation, Assessment, Valuation and Impact Pathways.*

When asked about the future of sustainability and the sustainability professional, Isidora Diaz Heredia, Chief Sustainability and HSE Officer, Parques Reunidos advises, "in the future sustainability will be 100% integrated into a company's strategy. The Chief Sustainability Officer will become the Chief Strategy Officer" (Heredia and Seabrook 2023). This second principle of the *Framework* ensures sustainability-related considerations are fully integrated into business decisions. Table 8.7 can assist with identifying where these decision-making considerations may exist within a company.

Sustainability Assessment and Implementation Tools

The sustainability assessment and implementation tools introduced in Table 8.8 with more detail in specific chapters noted parenthetically. They are part of the overall Integrated Sustainability Business Model, as shown in

Table 8.7 Integrating sustainability considerations into business decisions.

Integrating stainability considerations into business decisions such as:
• New product and service development and delivery
• Energy management
• Carbon emission reductions
• Worker health and safety
• Value chain transparency
• Governance
• Disclosure and reporting
• Biodiversity
• Water usage
• Deforestation and regeneration
• Public health impacts

Table 8.8 Sustainability assessment and implementation tools.

1) Frameworks and Applicability Analysis (Chapter 5)
2) Maturity Analysis (Chapter 9)
3) The Integrated Impacts, Risks, and Opportunities Management Cycle (Chapter 9)
4) The Sustainability (ESG) Strategy Development and Action Planning (Chapter 9)
5) Stakeholder Mapping (Chapter 10)
6) Benchmarking (Chapter 11)
7) Materiality Analysis (Chapter 12)
8) Value Assessment and Mapping (Chapter 13)

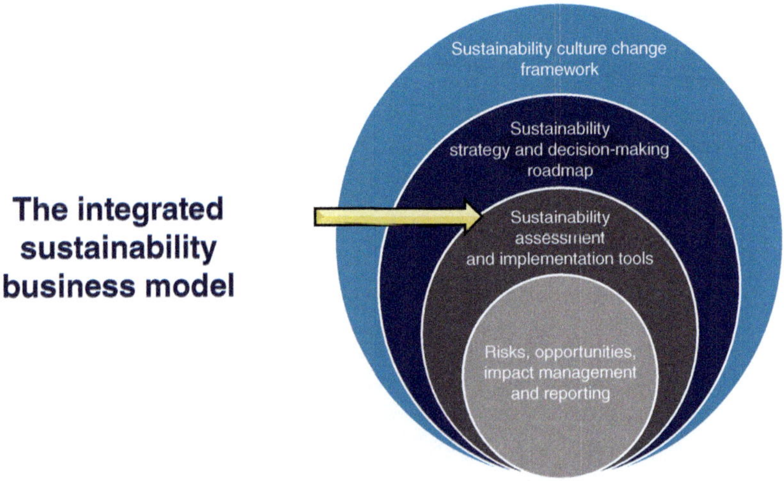

Figure 8.5 The integrated sustainability business model: sustainability assessment and implementation tools. Source: Global Solutions, Inc. and Seabrook (2024a, 2024b).

Figure 8.5. Inputs can assist in determining a company's current state of sustainability maturity, as well as provide the business intelligence needed to determine what an ideal future state of sustainability maturity might look like. The gap in this current and ideal future state of maturity will ultimately inform decisions on an aligned and integrated sustainability business strategy and action plan.

Sustainability Maturity Analysis

The Seabrook Sustainability and ESG Maturity Curve in Figure 8.6 is an assessment tool to assist with identifying the current state of a company's sustainability maturity journey. Using this tool to conduct a sustainability

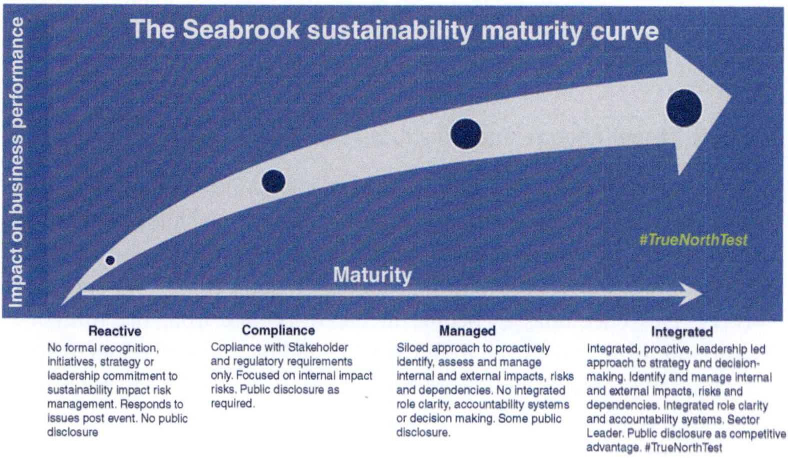

Figure 8.6 The Seabrook sustainability and ESG maturity curve. Source: Global Solutions, Inc. and Seabrook (2023).

maturity analysis further identifies organizational gaps in current and future state expectations, priorities, and decision-making. These organizational gaps provide important business intelligence, which in turn informs business strategy, action planning, and resource allocation to integrate sustainability material risk and impact management into business planning. A deeper discussion on the individual elements of the Maturity curve is found in Chapter 9 *Sustainability Strategy and Decision-Making Roadmap*.

Stakeholder Mapping, Materiality and Competitor Analysis

Identifying a company's stakeholders and mapping their sustainability expectations to the business, anticipates the implications for sustainability in strategy development and action planning. Stakeholder engagement is also implicit in both a materiality and competitor analysis. (Competitor Analysis is most often completed as part of Benchmarking. See Chapter 11 *Benchmarking* for more information on Benchmarking.) All three of these assessment and implementation tools provide data that informs and supports strategy and implementation prioritization and decision-making. A materiality analysis assists with identifying and prioritizing a company's key material risks, opportunities, and impacts. The competitor analysis helps align, focus, and prioritize what is important.

The *Sustainability Strategy and Decision-Making Roadmap* is a "how to," tool, integral to the integration of sustainability into the business. It provides guidance on integrating sustainability-related identification, assessment, prioritization, measurement, strategy development, and action planning processes. This Roadmap, and its applicability and use, is further discussed in Chapter 9.

Principle Three: Engage and Principle Four: Empower

In this author's experience with fortune 500 companies, "People are at the core of every thriving business and their ability to contribute is influenced by their health, safety and wellbeing" (Global Solutions, Inc. and Seabrook 2021, Seabrook 2023). People in this context include employees and other workers. This encompasses people in the workforce from the front line worker and team leads through to the CEO and board of directors or trustees. Other workers can include contract workers, value chain workers, volunteers, apprentices, and interns.

When starting any new initiative, engaging, and empowering the people involved increases the likelihood of success. In their collective experience, Dave Stangis and Katherine Valvoda Smith advise in their book, *The Executive's Guide to 21st Century Corporate Citizenship,* "the most vital (stakeholder) group is your company's employees. In our experience this key relationship is often under-leveraged" (Stangis and Valvoda Smith 2017). This relationship is a company's human and social capital. For more on the value of engagement and empowerment, see Chapter 4 *Theory of Change* for details on how Paul O'Neill, former CEO of Alcoa, leveraged employee safety to highlight the value of people and their "discretionary energy." Alcoa's cultural shift and increase in human and social capital was based on the premise of caring about employees and their safety and health, it was the lever or driver for change that improved overall performance for the company (BC Worker Safety 2014).

Engaging people increases their ownership, involvement, and commitment to the company and its initiatives, such as sustainability. This engagement influences the acceleration and integration of an initiative's implementation (Reeves 2023). Successful initiatives need supporters who influence others both formally and informally, with or without authority or accountability for the initiative. New initiatives take time to be embraced by individuals and integrated into the business. Whether the initiative is piloting an energy reduction plan for a warehousing operation or conducting a broader stakeholder mapping exercise, working with influencers such as internal corporate communications, investor relations, front line workers or procurement departments, these principles of engaging and empowering apply.

When thinking about collaboration, Sue Krzastek, a seasoned human resources professional in the healthcare and other industry sectors, shares, "you never know who may have critical information that could have a dramatic impact on safety, quality or sustainability. It is our job as leaders to

Table 8.9 The interconnection of engagement and empowerment.

Engagement without empowerment is futile. When stakeholders are empowered to contribute and believe their contributions will be considered and embraced in decision-making, they are more likely to be engaged and execute the decisions. This applies to all decisions, including those that are Sustainability-related.

find this information by making employees feel safe to share their thoughts" (Krzastek and Seabrook 2024).

Table 8.9 provides additional insight into the relationship between engaging and empowering people at all levels of a company.

For the EHS and sustainability professional, collaboration with internal stakeholders is an essential element of sustainability success. Not only does each internal stakeholder have influence within an organization, but they also hold the key to data and business intelligence. Table 8.10 identifies potential internal sustainability allies in a company and their role in sustainability. When asked which internal stakeholders the EHS and Sustainability professional should get to know to be successful, Mike Wallace suggests, "finance and procurement, sales and marketing, legal,

Table 8.10 Internal sustainability allies.

Departments	Role in sustainability
1) Finance	• Follow the money to understand the business. • There is financial detail on what is bought to operate the business and by what departments (i.e. spend on computers, tons of iron ore, air travel, renewable and nonrenewable energy, etc.)
2) Procurement	• Tracks purchases and activity generating CO_2e and contributing to the company's carbon footprint. • Assists with disclosure reporting: Voluntary and mandatory, customer requirements, and rating and ranking agencies (i.e. CDP, EcoVadis, MSCI, etc.)
3) Legal	• Monitors sustainability statements by CEO, board of directors, sales, and other company leaders. • Also monitors responses to requests for proposals (RFPs), assuring sustainability statements are accurate (i.e. net zero claims are true)

(*Continued*)

Table 8.10 (Continued)

Departments	Role in sustainability
4) Sales and marketing	• Understands the market and competition • Can identify significant customers and their sustainability expectations and requirements. • Can assist with positioning internal and external sustainability expectations
5) Investor relations	• If the company is publicly traded, get to know the investor relations officer. • Educate them on sustainability. • They produce the annual report and sustainability performance is part of the company's story.
6) Board of directors	• Get to know your board directly or indirectly through the corporate secretary, typically a role of general council. • Identify directors who can help drive company sustainability strategy and initiatives. • Understanding each board member's background such as their industry sector or other company board membership will provide a perspective on their sustainability fluency, maturity, and why it is a business imperative for the company.

Source: Adapted from Wallace and Seabrook (2023).

investor relations (if your company is publicly traded), and the board of directors" (Wallace and Seabrook 2023).

Helen Slinger, Executive Director, Accounting for Sustainability (A4S) recounts, "A4S's Theory of Change for the finance and accounting profession is about inspiring, transforming and scaling up financial professionals and financial teams within organizations. Inspiring them to understand the role they can play and why they (their role) are important. The next step is transforming the way they make decisions by helping them understand and action what they can do and why they should." It is about transforming the financial system (Slinger and Seabrook 2023). Finding ways to engage and empower sustainability-related stakeholders is key to optimal decision-making and sustainability alignment, integration, and performance.

When asked about collaboration and the future, Slinger suggests, "we are in a really different place today than we were a couple of years ago. The momentum around sustainability reporting make it more likely than ever that company finance teams are going to come knocking on the doors of their sustainability colleagues. This is also true in the area of sustainability assurance" (Slinger and Seabrook 2023). This requires breaking down silos within an organization, seeking collaborative solutions, and understanding how to execute on these 4 (four) principles of the *Sustainability Culture Change Framework*.

Summary

In this chapter, we have identified the key areas of alignment, integration, empowerment, and engagement for effective transformational culture change for long-term, sustainable value creation for all stakeholders. Shifts in accounting for finance, social and environmental factors, stakeholder expectations, and mandatory sustainability reporting requirements globally are driving this change.

In the next chapter, *The Sustainability Strategy and Decision Making Roadmap (Roadmap)* will be introduced. It is practical, relevant implementation guidance, which drills down into the processes for operationalizing sustainability. This *Roadmap* incorporates sustainability-related assessment tools, strategy development and action planning, including transparency, accountability, and reporting. The *Roadmap* guides company decision-making while continuing to create value for all stakeholders.

References

BC Worker Safety (2014). Tough decisions: Fareed Zakaria: CNN GPS special. https://www.youtube.com/watch?v=56a3-Sc65M8 (accessed 31 March 2024).

Brandon, C. and Seabrook, K.A. (8 September 2023). Interview with Chet Brandon, Corporate Director of Health, Safety & Environmental Protection, GrafTech International Ltd. Brandon, Chet (interviewee) and Seabrook Kathy A. (interviewer).

Capitals Coalition (2021). Social and human capital protocol primer. https://capitalscoalition.org/wp-content/uploads/2021/12/Social-Human-Capital-Protocol-A-Primer-for-Business.pdf (accessed 17 August 2023).

Cooke, M. and Seabrook, K.A. (17 May 2023). Interview with Michael Cook, Chief Sustainability Officer. Michael Cooke (interviewee) and Seabrook Kathy A. (interviewer).

Global Solutions, Inc. and Seabrook, K.A. (2021). Emerging issues impacting safety and health. “Connecting the Dots” guest lecturer at the college of the Mainland Gulf Coast Safety Institute (18 June 2021).

Global Solutions, Inc. and Seabrook, K.A. (2022a). #TrueNorth: EHS-ESG - are you and your company future fit? Seminar presented at the American Society of Safety Professionals Educational Event (24 October 2022).

Global Solutions, Inc. and Seabrook, K.A. (2022b). The EHS professionals expanding role in human capital, sustainability & ESG presented at the American Society of Safety Professionals webinar series (8 December 2022).

Global Solutions, Inc. and Seabrook, K.A. (2022c). Connecting the dots: sustainability, ESG, human capital and the safety professional presented at the Board of Certified Safety Professionals Global Learning Summit. Virtual (11 May 2022).

Global Solutions, Inc. and Seabrook, K.A. (2024a). ESG/sustainability: strategy, road mapping and action planning one-day seminar presented at Safety 2024 Pre-conference, American Society of Safety Professionals Annual Professional Development Conference (6 June 2024).

Global Solutions, Inc. and Seabrook, K.A. (2024b). The fundamentals of ESG and sustainability one day seminar presented at the American Society of Safety Professionals Safety Focus 2024 (22 February 2024).

Global Solutions, Inc. and Seabrook, K.A. (2024c). Connecting the dots: the value of people at work and their health, safety and well-being. *3 May 2024 Keynote Presentation at the 34th International Congress on Occupational Health*, Marrakesh, Morocco.

Heifetz, R.A. and Linsky, M. (2002). *Leadership on the Line: Staying Alive Through the Dangers of Leading*. Boston: Harvard Business School Publishing.

Heredia, I. and Seabrook, K.A. (19 and 26 June 2023). Interviews with Isidora Diaz Heredia, Chief Sustainability and HSE Officer, Parques Reunidos. Isidora Diaz Heredia (interviewee) and Seabrook Kathy A. (interviewer).

Krzastek, S. and Seabrook, K.A. (28 June 2024). Interview with Sue Krzastek, SPHR, SHRM-SCP, CEBS, MBA, Human Resources professional. Krzastek, Sue (interviewee) and Seabrook, Kathy A. (interviewer).

Reeves, T. (2023). Principles and practices of organizational change in the context of safety and culture. In: *Safety Professionals Handbook* (ed. J. Haight), 1107–1119. American Society of Safety Professionals.

Seabrook, K.A. (2023). Ten strategies for global safety and health management. In: Safety Professionals Handbook. (ed, J. Haight), pg. 1152. American Society of Safety Professionals.

Seabrook, K.A. and Reeves, T. (2018). *Safety Leadership and Professional Development* (ed. R. Olawoyin and D.C. Hill). American Society of Safety Professionals.

Slinger, H. and Seabrook, K.A. (2023). 25 September 2023 Interview with Helen Slinger, Executive Direct, Accounting for Sustainability. Slinger, Helen (interviewee) and Seabrook, Kathy A. (interviewer).

Stangis, D. and Seabrook, K.A. (2023). 18 July 2023 Interview with Dave Stangis, Chief Sustainability Officer. Stangis, Dave (interviewee) and Seabrook, Kathy A. (interviewer).

Stangis, D. and Valvoda Smith, K. (2017). *The Executive's Guide to 21st Century Corporate Citizenship*. Bingley, UK: Emerald Publishing Limited.

Staves, M. and Seabrook, K.A. (18 May 2023). Interview with Malcolm Staves, FIChemE, Global Vice President Health and Safety, L'Oréal (interviewee) and Seabrook Kathy A. (interviewer).

US Business Roundtable (2019). Statement on purpose of a corporation. https://system.businessroundtable.org/app/uploads/sites/5/2023/02/WSJ_BRT_POC_Ad.pdf (accessed 4 January 2024).

Wallace, Mike and Seabrook, Kathy A. (2023). 4 September 2023 Interview with Mike Wallace, Chief Carbon Officer. Wallace, Mike (interviewee) and Seabrook, Kathy A. (interviewer).

9

The Sustainability Strategy and Decision-Making Roadmap

Maximizing Impact and Value Through Strategy, Decision-Making, and Stakeholder Engagement

Following the Sustainability Culture Change Framework, in the *Integrated Sustainability Business Model (Model)* (Figure 9.1), which was described in Chapter 8, the next level in the *Model* is the *Sustainability Strategy and Decision-making Roadmap (Roadmap)*, highlighted with an arrow in Figure 9.1. The *Sustainability Assessment and Implementation Tools*, in Table 9.2 are also in *the Model* and support the Roadmap process.

Let us begin to unpack the components of the *Roadmap*, keeping in mind a company's impact and risk can be positive or negative. As highlighted in Table 9.1, the objective of sustainability-related initiatives is to enhance a company's positive impacts and opportunities, while mitigating negative risks to and impacts and dependencies on people, society and the environment, business, and economy.

The *Roadmap* presented in Figure 9.2 provides a sequenced, "how to," approach for integrating sustainability into business strategy, decision-making, and action planning. The *Roadmap's* purpose is to facilitate a shared organizational understanding of sustainability-related value and positive impact for a company and its stakeholders. Using the techniques of assessment, goal setting, measuring, monitoring, and improvement, the *Roadmap* guides users through a process of self-awareness on a current state to a newly identified ideal future state of sustainability maturity. The gap between a company's current and future state of sustainability becomes input or output, informing decision-making, business integration, prioritization, strategy development, action planning, value creation, and company

Connecting the Dots: Environment, Health, Safety, and Sustainability,
First Edition. Kathy A. Seabrook and Tanis J. Marquette.

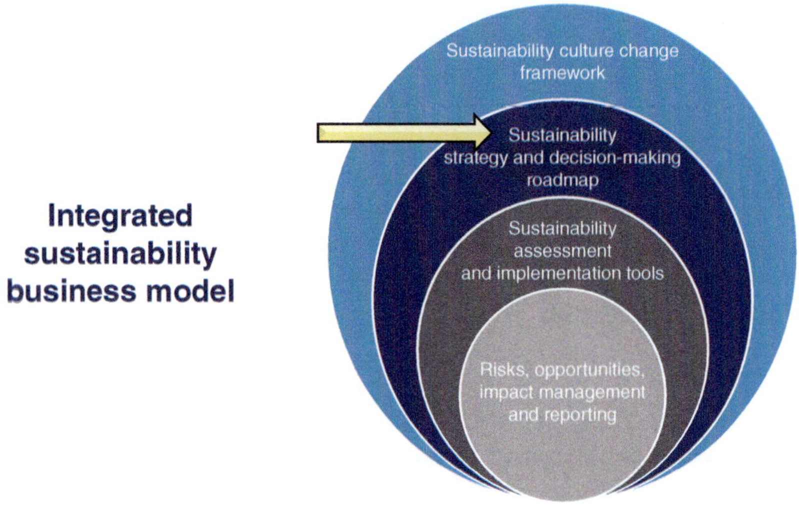

Figure 9.1 Integrated sustainability business model: sustainability strategy and decision-making roadmap. Source: Global Solutions, Inc. and Seabrook (2024).

Table 9.1 The objective of sustainability.

The objective of sustainability is to enhance a company's positive impacts and opportunities, while mitigating negative impact and dependency risks.

Source: Global Solutions Inc. and Seabrook (2024).

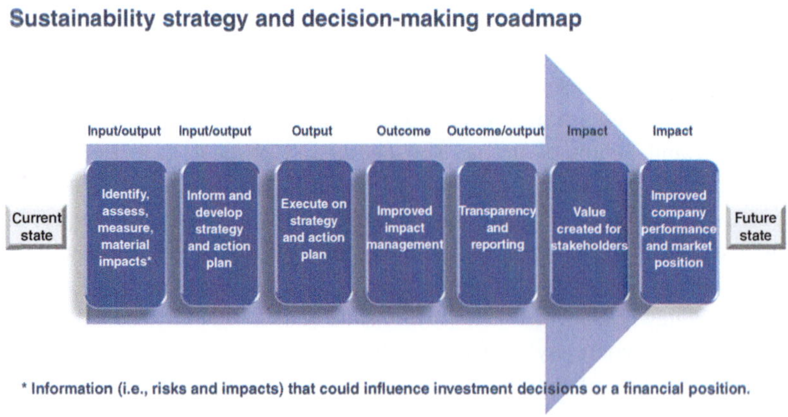

Figure 9.2 Sustainability strategy and decision-making roadmap. Source: Global Solutions, Inc. and Seabrook (2024)

performance. As discussed in Chapter 8 *Sustainability Culture Change Framework,* integrating sustainability into the business, its strategy, and operations is key to creating long-term value, positive impact, brand integrity, sustainability, and company performance.

The *Roadmap* relies on the *Sustainability Assessment and Implementation Tools* provided throughout this book and referenced in Figure 9.2. These tools enable a company to determine its baseline, its current state of sustainability-related maturity, and performance (See Figure 9.3). Using these sustainability assessment and implementation tools, the *Roadmap* begins with identifying, assessing, and prioritizing a company's internal and external material risks, impacts, and dependencies. More on impacts and dependencies is found in Chapter 4 *Theory of Change*. An example of

Table 9.2 Sustainability assessment and implementation tools.

1) Frameworks and Applicability Analysis (Chapter 5)
2) Maturity Analysis (Chapter 8)
3) The Integrated Impacts, Risks, and Opportunities Management Cycle (Chapter 9)
4) The Sustainability Strategy Development and Action Planning (Chapter 9)
5) Stakeholder Mapping (Chapter 10)
6) Benchmarking (Chapter 11)
7) Materiality Analysis (Chapter 12)
8) Value Assessment (Chapter 13)

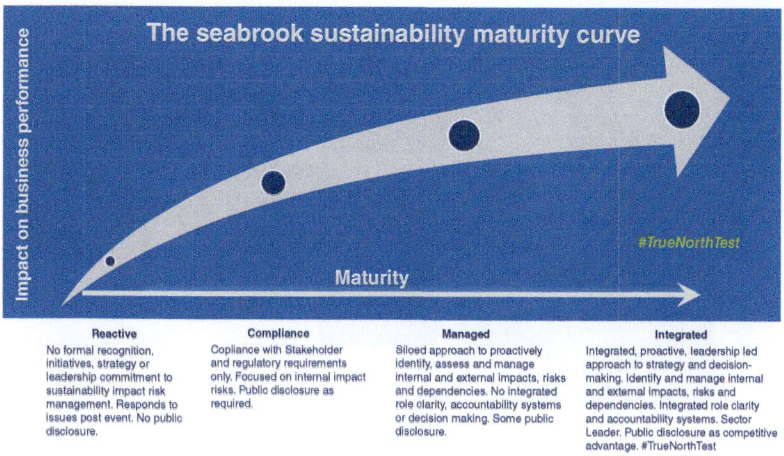

Figure 9.3 The Seabrook sustainability maturity curve. Source: Global Solutions, Inc. and Seabrook (2024).

material climate-related risk could be the impact of flooding or unusually hot temperatures on a company, its operations, workers, and value chain integrity. A company's worker safety and health or product safety risk and its impact on people, customers, suppliers, supplier's workers, communities, regulators, the environment, economy, and society are other examples of potential material risks, impacts, and dependencies. Chapter 12 *Materiality* provides additional information and examples of materiality, risks, and opportunities.

The first step in the *Roadmap* is aggregating data as collective input to determine a company's current state of sustainability. This informs sustainability-related strategy development and action planning. Data aggregation includes the efficacy of current risk, impact, and dependency management. The *Integrated Risks and Opportunities (Impacts) Management Cycle* in (Figure 9.4) can provide guidance on a risk-based management systems approach to the identification, assessment, measurement, monitoring, and continual improvement of material risks, impacts, and dependencies. It also assists with identifying opportunities and maximizing value for stakeholders through effective risk and impact management, which can ultimately influence a company's overall performance and market position. This is important business intelligence for decision-making by those leading and managing sustainability in a company.

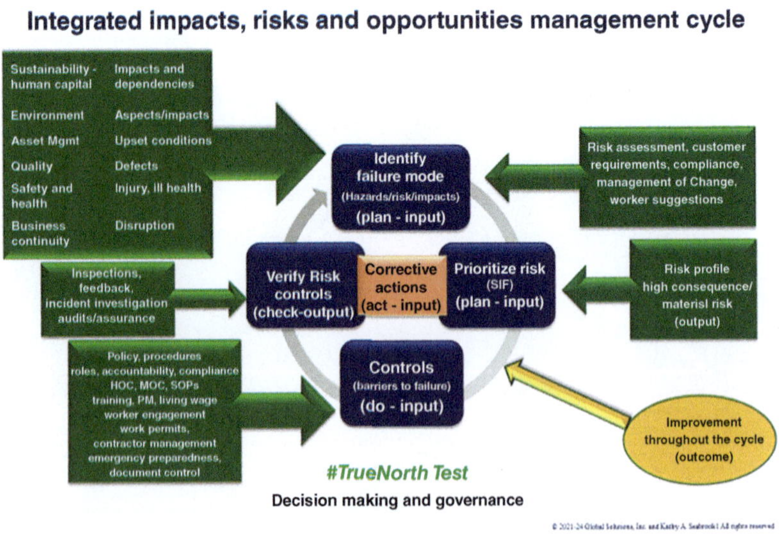

Figure 9.4 Integrated Impacts, risks, and Opportunities management cycle. Source: Global Solutions, Inc. Seabrook (2024).

Leading with Impact: Accessing the Current State of Sustainability

No matter where a company finds itself in the landscape of sustainability, whether it is in the earliest stages of self-discovery, identifying gaps and blind spots, or navigating new national and international mandatory and voluntary standards, this series of figures, diagrams, assessment tools, and accompanying usage descriptions will offer guidance and support.

The *Roadmap* in Figure 9.2 is conceptually based on the Theory of Change, incorporating the notion of **inputs**, **outputs**, **outcomes,** and **impacts** as part of a change process. The Theory of Change and its applicability to valuation and this change process are discussed in more depth in Chapter 4 *Theory of Change* and Chapter 13 *Value: Creation, Assessment, Valuation, and Impact Pathways.*

While the *Roadmap* aligns with the Theory of Change approach, it is important to remember the overall objective of sustainability is to enhance a company's positive impacts and opportunities. These impacts result from a change in the way a company is currently operating and making decisions. Understanding the Theory of Change concepts of input, output, outcome, and impact, as well as their relationship to one another, optimizes impact, risk and dependency identification, assessment, prioritization, mitigation, and management. The result is a move towards positive impact and material risk reduction. This resulting positive impact, coupled with reduced risk from dependencies such as worker health and safety risk, creates value for a company and its stakeholders.

The Roadmap

The following sections of this chapter break down each component of the *Roadmap* (See Figure 9.2) to provide more clarity on implementation steps.

Identify, Assess, Manage Material Risks, and Resulting Impacts (Input, Output)

Identify, assess, measure, material impacts*

The Sustainability Assessment and Implementation Tools (SAITS) in Table 9.2 provide guidance on identifying, assessing, measuring, and managing a company's material impacts, risks and dependencies. The *Maturity Analysis and Integrated Risk Management Cycle* are described below.

Maturity Analysis (Input/Output)

The Seabrook Sustainability (ESG) Maturity Curve (Maturity Curve) in Figure 9.3 is a qualitative sustainability assessment tool to help identify the current state of a company's sustainability maturity by depicting how an optimal, mature, sustainable company integrates sustainability into its decision-making and business. This assessment tool tracks a company's maturity from its initial introduction to sustainability through to the full integration of sustainability into the purpose, culture, decision-making, strategy, and actions across the business. The following breaks down each maturity mode.

Reactive Mode

A company's sustainability journey typically begins in a reactive mode where a company has no formal recognition, initiatives, or leadership commitment to sustainability (ESG) related impact or material risk management. That is, a company in reactive mode addresses sustainability-related impact as it arises rather than proactively. There is no intentional identification or public disclosure of sustainability-specific material risk.

Compliance Mode

Compliance mode is the next level of sustainability maturity. In this mode, a company becomes aware of sustainability-related risk, impacts, and dependencies through stakeholder requirements and mandatory disclosure regulations. In this maturity level, a company is focused on meeting those requirements and regulations. Stakeholder requirements may include CDP disclosure, disclosure of greenhouse gas emissions to key customers, or human capital management strategy and outcomes disclosed to specific investors. Human capital management outcomes can include retention rates, compliance with human rights standards, living wages for employees, worker career development and health, safety, and well-being. Depending upon the stakeholder and company boundary, working conditions in the value chain may also be included.

Managed Mode

The next level of maturity in the curve is managed mode. In this mode, a company is at a more mature state of sustainability maturity by proactively managing sustainability-related impacts and dependencies. While these sustainability-related impacts and dependencies are identified and managed proactively, it is not done using an integrated approach throughout the company. Sustainability initiatives are siloed and their interconnectivity across the business and its functional groups is not recognized by decision makers, presenting a missed opportunity for an advantage over competitors and supporting a company's brand.

Integrated Mode

Integrated mode is the most mature state of a company's sustainability integration. This state of sustainability maturity exemplifies an organizational focus on #TrueNorth (see Chapter 8 *Beginning the Journey: The Integrated Sustainability Business Model and Sustainability Culture Change Framework*) with proactive sustainability initiatives, fully integrated into business decision-making. Sustainability initiatives are leadership led and are aligned with a clear governance structure, role clarity, and accountability systems. The company publicly reports its sustainability-related performance, and sustainability is seen as an advantage over competitors and is leveraged to enhance the company's brand.

Understanding where a company is currently positioned on the *Maturity Curve* enables a company to benchmark using comparable, consistent modes or levels of sustainability maturity. Gaps identified as a result of benchmarking inform decision-making on sustainability-related strategy and action planning, where objectives and targets are developed to close the gaps to operating at an ideal future state of sustainability performance as determined by the company. More on benchmarking with peer companies or competitors is found in Chapter 11 *Benchmarking* with a detailed description and methodologies for conducting a benchmarking exercise.

The *Maturity Curve* is an important tool enhancing a company's ability to assess organizational readiness and leadership alignment on the current and an ideal future state of sustainability performance. It provides decision-useful information for strategy, integration, and implementation.

The Integrated Risk Management Cycle (Input, Output, Outcome)

Figure 9.4 depicts the *Integrated Risk Management Cycle (the Cycle)*. It integrates interconnecting processes, creating a cycle or system to initiate change within a business using a risk, opportunity, impact and dependency management improvement approach. The objective of *The Cycle* is to mitigate risk and negative impacts and dependencies that can erode a company's value. A governance structure for decision-making is also imbedded in *the Cycle*. Companies that are familiar with International Standards Organization environment and occupational health and safety management system standards will recognize this risk-based approach using a plan-do-check-act sequencing and improvement methodology. It originated from Edwards Deming's work in the 1950s (ISO 2015 and 2021, ISO 2018 and 2022, Deming and American Society for Quality 2024).

When the conversation (in a company or with stakeholders) begins to form around risk, "there is a common understanding for financial and sustainability professionals. Both manage risks, and good risk management provides opportunities for an organization to be in a strong, sustainable business position creating long term shareholder value," Slinger recounts from the successful A4S company case studies published on their website (Slinger and Seabrook 2023). This common understanding of risk holds true for EHS professionals as well.

From a people sustainability perspective, the Global Reporting Initiative (GRI) 403 Occupational Health and Safety (OHS) disclosure standard recognizes a management system approach for measuring and reporting OHS-related sustainability performance (Global Reporting Initiative 2018). As a member of the 2018 GRI 403 project group, the author notes, "the GRI 403 standard places emphasis on reporting proactive measures of risk reduction such as proactively identifying impacts (work) with high consequence injury and ill health potential (includes fatalities) rather than just reporting on incidents that have already occurred" (International Labor Organization and Seabrook 2018). Risk identification is embedded in a risk management system, a set of processes that assist a company with proactively identifying, measuring, monitoring, and improving upon its risk, opportunities, and risk management, keeping in mind dependencies can also present a risk to a company. An OHS (risk) management system is a leading indicator of performance and is affirmed as an OHS disclosure in the GRI 403 OHS standard, it is also recognized in many corporate sustainablity strategies and reports.

The interconnected impacts, risks and opportunities management cycle processes in Figure 9.4 help a company identify, assess, prioritize, and mitigate its material risks, enhance opportunities, and reduce negative impacts and dependencies. *The Cycle* promotes implementing control measures or barriers to eliminate high to low consequence risks and impacts through a hierarchy of controls. It further verifies these controls are effective and if they are not, applying additional controls (i.e. corrective actions) to reduce risk by further improving upon the management of material risks and negative impacts and dependencies.

The selection of effective risk control measures or barriers should follow a recognized hierarchy of control methodology. For example, ISO 45001:2018 *Occupational health and safety management systems: Requirements with guidance for use* standard (confirmed in 2022) provides a globally recognized hierarchy of controls methodology. This hierarchy example is outlined in the standard's clause 8.1.2 on "Eliminating Hazards reducing OH&S Risks" (International Standards Organization 2018 and 2022).

An additional output and outcome of *the Cycle* is identifying opportunities. Opportunities for improvement and negative risk and impact reduction can result in potential, positive impacts to people, the environment, products, services, business finance and efficiencies, society, and the economy. Opportunities can come from a change in leadership decisions, work organization, raw materials, operational processes and procedures, organizational accountability structures and improvement in the management system processes. Improvement in the management system processes could include risk assessment, incident investigation and management of change processes.

As an example, *the Cycle* identifies a risk and proposes a change or control measure, also known as an impact driver (i.e. plan [Input] from the Plan-Do-Check-Act model), such as installing a ventilation system in a hazardous chemical processing area. The change is intended to reduce greenhouse gas air emissions as well as exposure of workers and local communities to hazardous chemicals that can result in ill health or death. The risk identified can result in an impact, the risk mitigation or change is the impact driver (see Chapter 4 *Theory of Change* impact pathways examples). Risk prioritization is the output of risk identification and assessment (Seabrook 2023). The proposed change or impact driver is then implemented (do [input]), emissions and exposure levels are measured and verified (check [output or outcome] from the Plan-Do-Check-Act model), and additional action is taken if the risk is still too high (act [input, output, outcome] from Plan-Do-Check-Act model). Risk reduction is the outcome of *the Cycle*, which impacts the environment and people at work, in communities and society.

Examples of positive impacts resulting from the management of material risk:

- Increased product safety, reducing potential injuries or ill health on customers and society as well as the integrity of the company's brand reputation.
- Reduced hazardous air and water emissions lead to better air and water quality, impacting the health and quality of life for people, workers, communities as well as reducing financial costs on the company's balance sheet.
- Reduced greenhouse gas emissions helping to stabilize severe weather patterns and effects on people, workers, nature, business, and society in the long term.

When asked about leadership and integrated decision-making, Sue Krzastek, an experienced US-based, human resources professional believes, "good leadership education is so important because it teaches us to step away from day-to-day activities, to actually look at the larger picture and how inter-related decisions are creating high quality (and safe) employee work-life and customer products (Krzastek and Seabrook 2024).

The impact and effectiveness of *the Cycle* outcomes are a measure of a company's #TrueNorth purpose and can be used as part of a #TrueNorthTest on how well a company is aligned with its #TrueNorth purpose (Global Solutions, Inc. and Seabrook 2024). See Chapter 8 *Beginning the Journey* for more information on #TrueNorth.

In order to better understand and implement *the Cycle*, its five processes are further described below. (Seabrook 2023).

Failure Mode (Input)

Identifying failure modes (Figure 9.5), including hazards, aspects, risks and resulting negative impacts, and dependencies, is the initial process in *the Cycle*. Examples of hazards, risks, and impacts are found in Table 9.3 and begin to demonstrate how *The Cycle* can assist in integrating the identification, assessment, and management of enterprise and sustainability-related risks. The ultimate goal of the tool is to provide a process for continually improving risk and negative impact management. Continual improvement creates opportunities for a business by valuing people and nature as part of its profit and loss reporting. More on this in Chapter 4 *Theory of Change* where impact pathways are discussed, and examples provided on the relationship between valuation and impact.

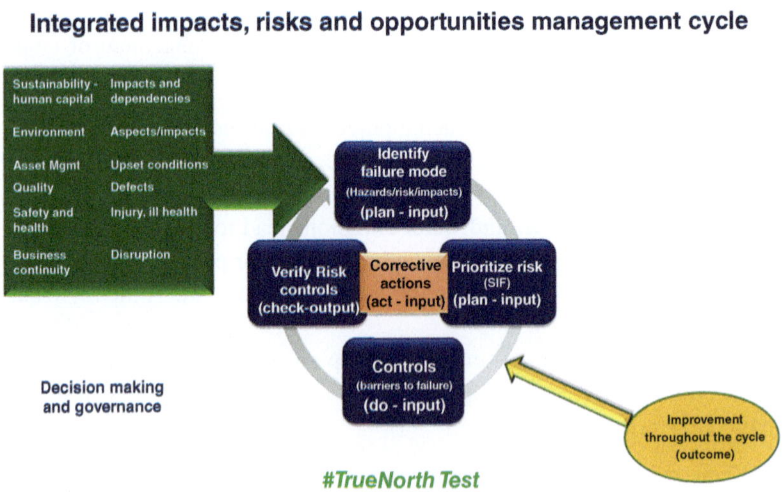

Figure 9.5 Integrated impacts, risks, and opportunities management cycle: identify failure mode I. Source: Global Solutions, Inc. and Seabrook (2024); adapted from the work of T.R. Reeves (2023).

Table 9.3 Integrated impacts, risks and opportunities (impacts) management framework: identifying failure mode.

Identifying failure modes (hazards, risks, impacts)	
Risks	**Failure mode**
Environmental	Aspects and impacts: spills, pollution, greenhouse gas emissions, water availability, deforestation, availability of other natural resources
Safety and health	Injuries and ill health
Asset management	Upset conditions and unscheduled equipment downtime
Quality	Product defects and customer complaints
Sustainability	Material risks, impacts and dependencies on people, business relationships, the environment, access to natural resources, water scarcity, climate risk (flooding, high temperatures, fires)
Business Continuity	Weather-related plant and production shutdowns; supply chain disruption and labor shortages

Source: Global Solutions, Inc. and Seabrook (2024).

Impact, risk, and hazard failure modes can be identified through a life cycle analysis, task-based job hazard analysis/risk assessment, Kaizens and other Six Sigma tools, and management of change processes. Actual incidents, maintenance logs, International Sustainability Standards Board IFRS materiality standards, industry best practice, worker suggestions, compliance obligations, customer requirements, and other stakeholder requirements are also sources to identify failure modes.

Assessing and Prioritizing Risks, Opportunities, Impacts, and Dependencies (Output)

Figure 9.6 highlights risk assessment and prioritization processes, the next step in *the Cycle*. Once the risk and impact failure modes are identified, they are prioritized based on significance and materiality. Table 9.4 provides some examples of prioritized risks and impacts to consider. Prioritization is a critical step. It enables a focused approach to resource allocation and mitigating and managing the most significant risks that have the greatest impact or upon which the company has the greatest dependency.

The results of a Materiality Analysis will assist in this prioritization process. Risk, impact, and materiality mapping also provide decision makers with a visual representation of the company's most significant and material risks by priority and impact. Figure 9.7 depicts an example of a materiality

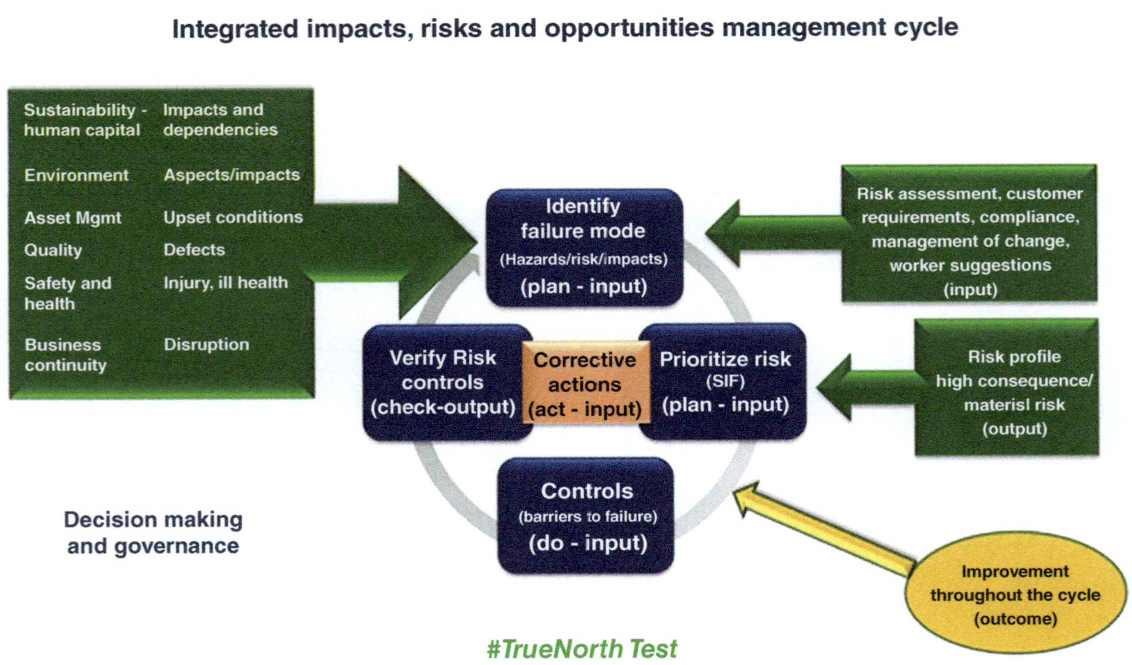

Figure 9.6 Integrated impacts, risk management cycle: Assessment and Prioritization. Source: Global Solutions, Inc. and Seabrook (2024); adapted from the work of T.R. Reeves (2023).

Table 9.4 Examples of prioritized risks and impacts.

- *Environmental* – The risk of a hazardous materials release polluting local aquifers, reducing freshwater availability for workers and communities (Impacts: human, social, and natural capital) and manufacturing operations. (Impacts: produced and financial capital.)
- *Safety and Health* – Potential and actual serious injuries, illness, and fatalities (SIIF) from an explosion risk due the improper storage and use of hazardous materials. (Impacts: human, social, natural, produced, and financial capital.)
- Asset Management – Downtime of boilers, key production equipment or the integrity of sprinkler systems or other fire suppression systems injuring workers, damaging property and products, and disrupting operations and fulfilling customer orders. (Impacts: human, social, produced, and financial capital.)
- *Quality* – Public health impacts due to the risk of defects in the efficacy of a pharmaceutical company's vaccines. (Impacts: human, social, produced, and financial capital.)
- *Sustainability* – (i) management and disclosure of greenhouse gas emissions; investor relations; severe weather risk on operations; reputational risk to company brand (Impact: human, social, produced, financial, and natural capital.) (ii) The composition of a corporate board impacting innovation and decision-making. (Impacts: human, social, produced, and financial capital.)
- *Business Continuity* – A company's dependency on a sole supplier of lithium for battery manufacturing.

map. Chapter 12 *Materiality* and Chapter 4 *The Theory of Change: Inputs, Outputs, Outcomes, and Impact* provide additional information and examples on how to conduct a materiality analysis, which provides critical business intelligence and informs decision-making. .

Once risks and impacts are identified and prioritized, this output is incorporated into the overall business strategy and action planning.

Barriers to Failure (Input)

Barriers to failure (Figure 9.8) reduce the level of a company's risk and negative impacts by providing controls or barriers that mitigate or eliminate risk and negative impacts. The hierarch of controls principles are applied to control/barrier decisions. Examples of barriers or risk controls are highlighted in Table 9.5.

Barrier Integrity (Output)

This process in *the Cycle* (Figure 9.9) verifies the barriers, also known as risk controls, are in place and effective. Processes to ensure barrier

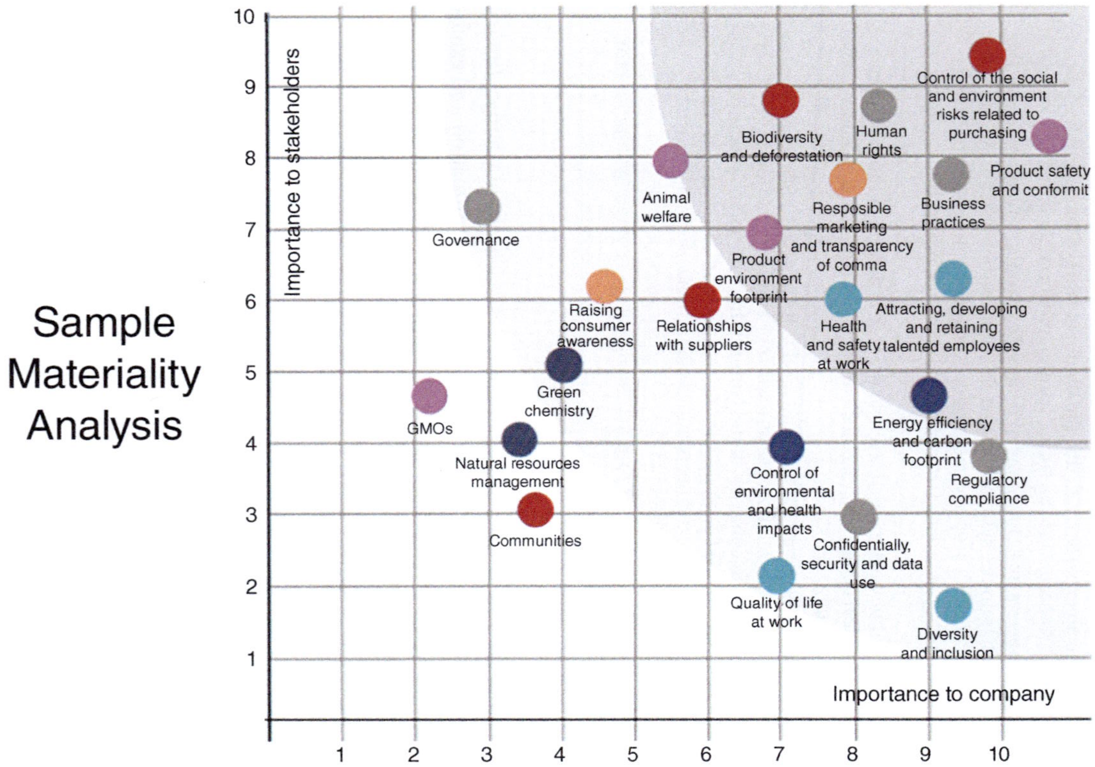

Figure 9.7 Sample materiality analysis. Source: (Global Solutions, Inc. and Seabrook 2024)

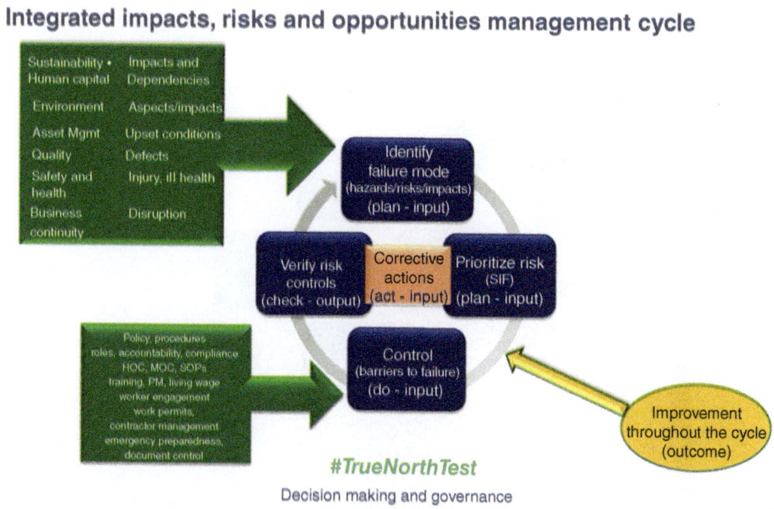

Figure 9.8 Integrated impacts, risks, and opportunities management cycle: barriers to failure. Source: Seabrook (2024)/Global Solutions, Inc

Table 9.5 Examples of barriers (risk controls).

Barriers (risk controls)
• Policies and procedures, • assurance of governance and accountability systems, • integrated risk management assurance (*the Cycle*), • management of change processes, • consistent hierarchy of controls decision-making, • standard operating procedures, • leadership commitment and engagement, • worker participation and engagement, • living wage assurance, • compliance assurance processes, • permits to work, fire protection systems, • personal protective equipment, • ventilation systems, • grounding and bonding systems for flammable liquids, • preventative maintenance programs, • worker suggestions, • training and development, • document control, • emergency preparedness, and • contractor management

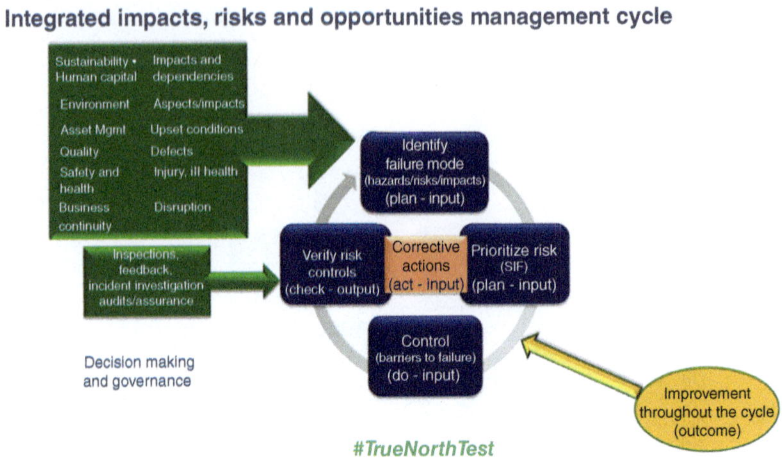

Figure 9.9 Integrated impacts, risks, and opportunities management cycle: verify barriers/Risk Controls. Source: Seabrook (2024)/Global Solutions, Inc.

integrity can include inspections, incident investigations, audits and assurance mechanisms, feedback from stakeholders, trend lines in unscheduled equipment downtime, customer complaints, or investor impact management inquiries.

When there are operational changes, leadership and policy changes, an incident occurs, or an inspection or audit finding is identified, a corrective action is generated. This is especially true for all potential or actual material risks such as:

- product quality,
- environment and nature,
- workers, customers,
- relationships with key stakeholders (i.e. investors and regulators),
- value chain disruption,
- damage to physical assets,
- continuity of business, and
- resilience.

Corrective actions bridge the gap between the initial implementation of barriers and the need for additional barriers in the event of failure. This gap is known as residual risk, where additional control measures are required to maintain a barrier to failure.

Continual Improvement (Outcome)

Corrective actions are inputs that lead to potential impacts and improvement. According to ISO 45001 *Occupational health and safety management systems: Requirements with guidance for use*, a corrective action is an "action to eliminate the cause(s) of a nonconformity or an incident and to prevent recurrence" (International Standards Organization 2018). Corrective actions lead to improvement (Figure 9.10), which in turn continues the cycle again. Improvement is an outcome of *the Cycle*. For example, incidents lead to new risk assessments with identification, prioritization, implementation of new or additional control barriers, verification of barrier integrity, and the improvement cycle continues. In the context of sustainability and impact, a corrective action is a catalyst for change, ultimately driving impact (Table 9.6).

The Cycle in its entirety is a tool that enables a company to demonstrate to stakeholders its integrated, systematic governance approach to decision-making and how well it is continually improving upon managing material risks and impacts. EHS professionals are very familiar with this risk-based approach to managing environmental, health, and safety risks using ISO management system standards. Additional assurance can include third-party validation of these management systems through certification.

Depending upon its level maturity, a company can apply this risk-based management approach in a siloed or integrated way. If integrated into strategy and across departments, divisions, or business units, there are

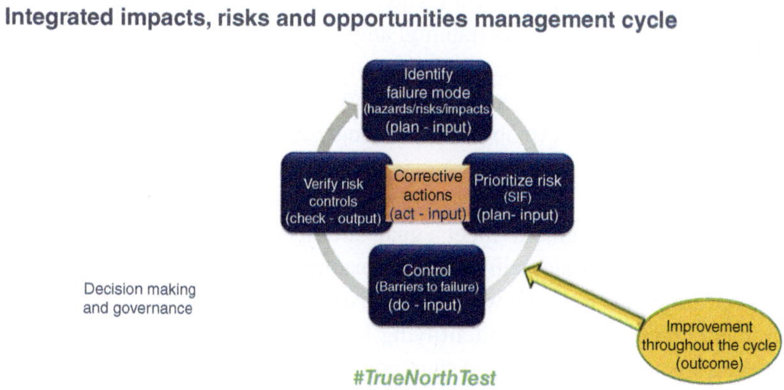

Figure 9.10 Integrated impacts, risks and opportunities (impacts) management cycle: improvement. Source: Seabrook (2024)/Global Solutions, Inc.

Table 9.6 The Relationship Between a Corrective Action and Impact.

A corrective action is an input, a catalyst for change that can drive impact.

consistent role clarity and accountability systems for identifying and managing stakeholder expectations and requirements, internal and external risks, opportunities, impacts, and dependencies. A siloed approach is a missed opportunity. Systemically integrating risk and impact management into the business strategy and operations creates value for both the company and its stakeholders.

By applying the same principles and methodologies to reduce sustainability-related risk and impacts across an organization and its business functions, a company's material risk and negative impacts are reduced in an integrated way. Opportunities for improved decision-making, operational efficiencies, employee retention, reduced injuries and ill health, productivity gains, cost reductions, environmental impacts and dependencies, and the well-being of workers and other stakeholders are also realized. These are just some of the opportunities that can create value for a company.

Developing Strategy and Action Planning (Input)

Inform and develop strategy and action plan

Outputs and outcomes from the Sustainability Assessment and Implementation Tools (SAITs) listed in Table 9.2 inform decision-making by providing input to sustainability strategy development and action planning. The intent of the strategy is to set out the vision and expectations of the company's future state of sustainability, aligned with its #TrueNorth. A 50% greenhouse gas emissions or solid waste reduction by 20XX + 5 are examples of goals incorporated in the company strategy. Companies typically develop short and longer-term goals, which can extend from 1 to 10 years.

Action planning sets out specific objectives, targets, accountability, authority and a timeline to meet a goal, such as the 50% reduction goal for greenhouse gas emissions in the above example. The action plan tactically sets out how the company will accomplish its goals. It incorporates measurement toward the goals, identifying key performance indicators, while integrating an accountability system and timeline to ensure objectives, targets, and overall sustainability performance expectations are met.

The *Maturity Analysis* (Figure 9.3) is also found in the list of SAITs in Table 9.2. This analysis provides guidance on a company's current state of sustainability integration. This becomes the starting point, the initial state of a company's sustainability journey. Comparing current state to the company's maturity aspirations provides guidance and input for developing a sustainability strategy and resulting action plan. Referring back to the maturity analysis, if a company is in "reactive mode, " at the beginning of their sustainability journey, one of their first steps should be to identify their sustainability-related stakeholders and better understand their sustainability-related expectations. Maturity Analysis is discussed in depth at the beginning of this chapter.

Stakeholder mapping is a SAIT and described in more depth in Chapter 10 *Stakeholder and Stakeholder Mapping*. It is a process that maps stakeholders affected by the company's sustainability performance to those same stakeholder's expectations of that performance. Stakeholders can include employees, workers in the value chain, customers, communities, suppliers, investors, regulators, rating agencies, policy makers, insurers, and nongovernmental organizations (NGOs).

The Applicability Analysis is another SAIT (noted in Table 9.2). It helps a company determine which mandatory and voluntary disclosure and other sustainability-related standards and frameworks apply to the company's business operations. The analysis considers the applicability of mandatory and voluntary country-specific and international standards such as human capital and climate change regulations, the European Corporate Sustainability Reporting Disclosure directive, or the International Sustainability Standards Board (ISSB) International Financial Reporting Standards. Country-specific environmental, occupational safety and health, and other labor regulations should also be considered. Stakeholder requirements such as CDP disclosure scoring of carbon emissions, water, or deforestation are also included in this applicability analysis. See Chapter 5 *Frameworks* for more detailed information on conducting an analysis.

The outputs and outcomes of these SAITs are then incorporated as inputs to the strategy and action planning. Once the strategy and action plan are agreed upon by key leaders and clear resource allocation and accountability expectations are integrated into the overall business strategy, the next step is to execute the strategy and action plan.

Executing a Strategy and an Action Plan (Output)

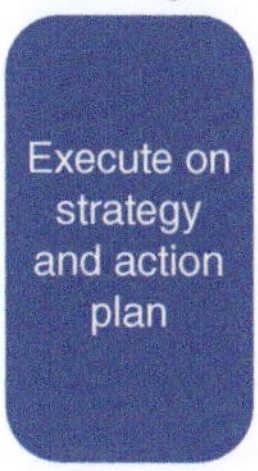

Executing a strategy and action plan aligns with the Sustainability Culture Change *Framework (Framework)* and its principles, practices, and processes for implementing a change such as embarking on a company-wide sustainability journey. The *Framework* is discussed in Chapter 8 *Beginning the Journey: The Integrated Sustainability Business Model and Sustainability Culture Change Framework*. Actively incorporating the *Framework* imbeds strategy and action planning into the business and engages and empowers decision makers at all levels of the company. This will have a significant influence on the success of implementation, ultimately resulting in better sustainability performance.

This step of the *Roadmap* is where goals, targets, roles, and responsibilities are identified, communicated, resources are allocated, and accountability systems are put in place. The following is also accomplished in this step:

- assurance processes for internal and third-party audits on material financial and sustainability reporting are in place,
- sustainability-related data gathering, trend analysis, and performance reporting,
- status on progress toward the Sustainable Development Goals and human rights issues,
- processes and technology to account for applicable value chain disclosure in company reporting,
- status on meeting the needs and expectations of stakeholders
- status on compliance with regulations, directives, and voluntary standards
- conformance with applicable international voluntary standards, including IFRS/ISSB, GRI, ISO, and industry-specific best practice.

The *Integrated Risks and Opportunities (Impacts) Management Cycle* (Figure 9.4) and impact pathway and value mapping (discussed in Chapter 4 *Theory of Change*) are also SAITs and can assist with active and ongoing identification, assessment, prioritization, management and improvement of material risks, opportunities, and resulting positive impacts. These inputs, outputs, outcomes, and impacts all influence decision-making and implementation across the company and its operations.

Improved Impact Management (Outcome)

Improving the management of impacts is an outcome that can lead to value creation for people, nature, society, the company, and economy.

The *Integrated Risks and Opportunities (Impacts) Management Cycle* (Figure 9.4), discussed earlier in this chapter, also incorporates an improvement process for impact management. There are other, general, industry, and company-specific processes and systems that can be used for process improvement. Lean, Six Sigma, Failure Modes and Effects Analysis, and Kaizens are improvement tools that take into account the Sustainability Culture Change *Framework* principles and practices, and also can be employed at this step in the *Roadmap*.

Transparency and Reporting (Output)

Stakeholders continue to seek transparency and disclosure from companies in order to make better decisions. These stakeholders include investors, customers, employees, suppliers, regulators, insurers, and rating agencies. For

Transparency and reporting

more information on transparency and reporting trends see Chapter 2 *Evolution of Sustainability*, Chapter 5 *Frameworks*, and Chapter 14 *Disclosures, Reporting, Greenwashing and Assurance*.

Public disclosure by companies is currently being driven by stakeholder's requirements and mandatory or statutory standards. The Applicability Analysis discussed in Chapter 5 *Frameworks*, assists and informs decisions on disclosure information to meet mandatory annual financial reporting requirements. The Corporate Sustainability Reporting Standard in Europe or the US Securities and Exchange Commission's human capital and climate change disclosure regulations require specific material disclosure information, including sustainability-related issues, in an annual financial report.

There are other drivers of public disclosure of sustainability-related information. Companies further along on their sustainability maturity journey understand the importance of proactively owning their sustainability narrative, typically published in sustainability or corporate responsibility reports. While not mandatory, this strategy provides decision-useful information for investors, employees, suppliers, and customers. This disclosure can ultimately impact access to capital and labor as well as increased product, service, and market opportunities.

Nike Inc.'s "Move to Zero" initiative targets carbon emissions and waste reduction. It is an example of a company further along in their sustainability maturity journey. One target in the initiative includes 100% waste diverted from landfills in Nike Inc.'s supply chain. This includes 80% or more of waste recycled back into Nike Inc. products (and other goods) by 2025 (Nike Inc. 2023). This is a good example of a company understanding their impact and proactively managing and publicly communicating that impact with stakeholders. In this case, positive social and environmental impact.

Finally, an important aspect of transparent disclosure is assurance. Assurance is third party validation of the data and other information publicly disclosed. For more details on assurance, see Chapter 14 *Disclosures, Reporting, Greenwashing, and Assurance*.

Value Created for Stakeholders (Impact)

Value created for stakeholders

Expectations of stakeholders identified in the *Stakeholder Mapping* exercise in the *Roadmap*, (Inform and Develop Strategy and Action Plan step), are linked to this VI. Value Created for Stakeholder' step. The Impact Pathways tool introduced earlier in this chapter and more fully in Chapter 13 *Value: Creation, Assessment, Valuation, and Impact Pathways*, assists with identifying impact drivers that create value for stakeholders such as workers and their families, communities as well as the business and its value chain, customers, the economy and society.

Stakeholder value, realized through impact identification mitigation and management, can be incorporated into the company's financial and sustainability performance reporting and communication mechanisms. Stakeholder value includes employee trust, loyalty and retention, suppliers' access to natural resources such as water and raw materials, customer trust and loyalty, business continuity and resilience, investor decision-making, and access to capital for acquisitions and other capital expenditures.

In the case of a pharmaceutical manufacture of insulin or heart medication, reducing single-source supply dependency and resulting disruption risk impacts better health outcomes and creates value for communities, families, and workers. Impact can also present as opportunities to enhance brand value and an advantage over the competition, creating value for the business, employees, suppliers, customers, investors, etc.

Improved Company Performance and Market Position (Impact)

Improved financial performance and market position is a goal of every company. Increasing stakeholder value can positively impact the overall performance of a company, including its market position. When customers require sustainability-related actions and performance, meeting those expectations could mean access to new and existing markets, thereby increasing market share. In some cases, not meeting customer requirements could bar the company from competing. An example is a customer requirement for CDP disclosure scores or loss time incident rates in a bid proposal. If the company does not report to CDP and does not have a score or has a high loss time incident rate, it may be excluded in the customer's bid decision-making process.

Positive sustainability-related impacts create value for a company in the area of worker safety and health. For example, managing workplace safety and health risks influences the overall well-being of employees and the likelihood they will be more fulfilled, productive, and innovative in their work. Employee well-being can also be influenced by opportunities for development, career advancement, engagement, a living wage, and decision-making empowerment (Capitals Coalition, L'Oreal 2022).

Ultimately, employee well-being impacts the level of employee trust, loyalty and retention, which increases the value of human capital (employees) to the company and can have a positive impact its valuation (Seabrook and Reeves 2018; Capitals Coalition 2019).

A company's sustainability performance and correlation to improved overall performance and market position influence decision-making and future business opportunities. Positive financial, human, social, and environmental impacts can create opportunities and the ability to create value by monetizing advantage over competitors in the market. This value creation is not a traditional methodology to account for the value of people and the environment in companies. It is a missed opportunity, one that could enhance profit and loss reporting.

Chapter 13 *Value: Creation, Assessment, Valuation and Impact Pathways* provides valuation examples using a value assessment, impact pathways approach to identifing value created by people, business relationships and the environment.

Case Study

According to Dr. Georgi Popov, professor, Occupation Risk and Safety Sciences, University of Central Missouri, "good risk management positively impacts a company's reputation and business continuity. When a company is prepared, it is more resilient" (Popov and Seabrook 2023). See the case study demonstrating this perspective (Table 9.7).

Conclusion

The *Sustainability Strategy and Decision-Making Roadmap* (Figure 9.2) is an implementation tool, supporting a company's sustainability-related journey. It provides guidance to identify gaps in a company's current sustainability initiatives. It can be implemented sequentially or as applicable, depending upon the company's sustainability maturity. Ultimately, the *Roadmap* provides business leaders with guidance on integrating sustainability into their business strategy, decision-making, and action planning. Implementing the *Roadmap* moves a company from its current to an intended ideal future state of sustainability maturity and performance.

Table 9.7 Case Study: Influencing decisions: connecting risk management and sustainability.

Reputation is important to a small or large company. In this example, a large US multinational company required their sites and supply chain to work towards certification to the ISO 45001 Occupational Health and Safety (OH&S) Management System standard over a 2.5 year period. The company also informed its suppliers there is an expection for them to move towards ISO 45001 certification. ISO risk based management system standards provide a critical framework for executing on sustainability and ESG. Implementing these standards can lead to: • improved reputation • reduced risk from less hazardous waste generation • better financial results due to illnesses and injuries and environmental impact risk reduction. • reduced insurance premiums • improved supply chain management • better governance through an audit process (OH&S, environmental, energy management, etc.) To drive this example home, the risks associated with the spray-painting operations at the company's local manufacturing site were assessed. The spray-painting operation used oil-based paint, benzene, and acetone. The employees were provided with face masks; however, they did not protect against exposure to organic compounds. Employees complained that their beer did not taste good, and their families did not like how they smelled after when they came home from working at the site. When the OH&S, environmental, community and sustainability risks and impacts were assessed, the following was found: 1. OH&S: Benzene and toluene presented a high health risk to workers; fire exposure, potential eye injuries due to chemincals use and management. These risk were not well controlled. 2. Environment: emissions from the chemicals presented a high risk and impact on the environment and community health. These risks were not well controlled.

(Continued)

Table 9.7 (Continued)

Although environmental and OH&S risk was high, it was not the primary influence to eliminate this risk. The reputational risk to the company's brand integrity and its sustainability over time drove the change. Standard practice was to exhaust the spray- painting emissions directly to the outside. Depending upon the wind direction and intensity, this did not cause problems because the area did not smell of chemicals all the time. This changed when the city decided to build a community park near the manufacturing site. When local citizens began to complain about the intermittent smell, it influenced the company's decision-making. They replaced the existing spray- painting operation with a power coating process. The new power coating process eliminated the use of benzene and acetone, reducing the OH&S, environmental, potential community health and reputational risk associated with the spray-painting operation.
This company scenario demonstrates the interconnectivity of brand integrity, economic impact,OH&S, environment, corporate social responsibility (CSR), and sustainability. It also reflects the value created by EHS, CSR and sustainability for a company's reputation, brand integrity, business continuity and ultimately financial position.

Source: Seabrook (2023)/Global Solutions, Inc.

References

Capitals Coalition (2019). Social and human capital protocol. https://capitalscoalition.org/capitals-approach/social-human-capital-protocol/ (accessed 8 August 2023).

Capitals Coalition, L'Oreal (2022). Valuing human capital in occupational health & safety. https://capitalscoalition.org/publication/valuing-human-capital-in-occupational-health-safety/ (accessed 17 August 2023)

Deming, W.E. and The American Society for Quality (2024). What is the plan-do-check-act (PDCA) cycle? https://asq.org/quality-resources/pdca-cycle (accessed 24 June 2024).

Global Reporting Initiative (2018). Topic standard: GRI 403: occupational health and safety 2018. https://www.globalreporting.org/how-to-use-the-gri-standards/gri-standards-english-language/ (accessed 7 November 2024).

International Labor Organization and Seabrook, K.A. (2018). Connecting the dots: occupational safety and health and business performance. Think Piece: Kathy Seabrook. World Day for Safety and Health at Work 2019 https://www.ilo.org/global/topics/safety-and-health-at-work/events-training/events-meetings/world-day-for-safety/33thinkpieces/WCMS_681587/lang--en/index.htm (accessed 5 March 2024).

International Organization for Standardization (ISO) (2015, confirmed 2021). ISO 14001:2015 Environmental management systems. Requirements with guidance for use. International Standards Organization.

International Organization for Standardization (ISO) (2018). ISO 45001:2018 Occupational health and safety management systems. Requirements with guidance for use. International Standards Organization. Reconfirmed in 2024.

Krzastek, S. and Seabrook, K.A. (28 June 2024). Interview with Sue Krzastek, SPHR, SHRM-SCP, CEBS, MBA, Human Resources professional. Krzastek, Sue (interviewee) and Seabrook, Kathy A. (interviewer).

Nike, Inc. (2023). Move to zero. https://www.nike.com/sustainability (accessed 31 December 2023).

Popov, Dr. G. and Seabrook, K.A. (2023). Interview with Georgi Popov, PhD, QEP, CSP, ARM, SMS, CMC, FAIHA, Professor, GPS Chair (Interim), Safety Sciences, University of Central Misssouri. Popov, Dr, Georgi (interviewee) and Seabrook, K.A. (interviewer)

Reeves, T. (2023). Principles and practices of organizational change in the context of safety and culture. In: *Safety Professionals Handbook* (ed. J. Haight), 1107–1119. American Society of Safety Professionals.

Seabrook, K.A. (2023). Ten strategies for global safety and health management. In: Safety Professionals Handbook. (ed, J. Haight), 1107–1119 and 1155. American Society of Safety Professionals.

Seabrook, K.A. (2024). The fundamentals of ESG and sustainability one day seminar presented at the American Society of Safety Professionals Safety Focus 2024 (22 February 2024).

Seabrook, K., A. and Reeves, T. (2018). *Safety Leadership and Professional Development* (ed. R. Olawoyin and D.C. Hill). American Society of Safety Professionals.

Slinger, H. and Seabrook, K.A. (2023). 25 September 2023 Interview with Helen Slinger, Executive Direct, Accounting for Sustainability. Slinger, Helen (interviewee) and Seabrook, Kathy A. (interviewer)

10

Stakeholders and Stakeholder Mapping

Introduction

Understanding your stakeholders is an important aspect of executing your sustainability strategy. We have written this section to provide a little something for everyone – we hope. The most wonderful aspect of sustainability is that it is a kaleidoscope of methodologies. We are sharing from our combined 60-plus years of experience in hopes of clearing hurdles for those that are just now in the throes of their careers or being introduced to sustainability; however, that does not at all indicate how you and your organization has proceeded was any less valuable.

From understanding what is important and unique to each stakeholder group and perhaps even within each stakeholder group to how to engage each group, and how often, stakeholders can hold the key to successful sustainability implementation Cerri (2024). In this chapter, we will identify each group, provide some examples of how you might engage stakeholders, and how organizations disclose their engagement strategies in generic terms. Upon conclusion of this chapter, you should be prepared to complete your own personal, organizational, or value chain stakeholder mapping, which can be used to inform your strategy. You may also find other value uses for stakeholder mapping beyond sustainability where communication and engagement pathways are critical to culture change success . . . (hint hint)!

What Is It?

An internet search of the definition for "stakeholder" returns "denoting a type of organization or system in which all the members or participants are

Connecting the Dots: Environment, Health, Safety, and Sustainability,
First Edition. Kathy A. Seabrook and Tanis J. Marquette.

seen as having an interest in its success" as defined by Oxford Oxford Learners Dictionaries (2024). Using that as the base of our understanding of stakeholders as it applies to sustainability, stakeholder groups are entities or collectives of groups in which members have a common interest in successful outcomes under the umbrella of sustainability. What you have no doubt discovered in your experiences thus far and in reading this book, not all stakeholders see sustainability through the same lens and therefore also do not see success in the same way either. However, despite that small challenge, we will make every effort to bring you closer to that realization. Think about sustainability, stakeholders, and success like a kaleidoscope. Every turn of the kaleidoscope garners a new and different wonder – each unique and breath-taking in its own way, described very similarly and simultaneously not at all similar and yet each is a success. Sustainability engagement with stakeholder groups can be much like kaleidoscope dances. Enjoy each one for their unique and colorful journey. Let us jump in with the different types of stakeholder groups you might be working with at different times through your sustainability journey and how those engagements culminate in the stakeholder map

Stakeholder Groups

Workers, Labor Unions, Contractors
Leadership, Board Members, Owners
Investors, Shareholders, Financial Institutions
ESG Raters, ESG Rankers
Certification Bodies, Assurance Providers
Regulators, Non-Governmental Agencies, Voluntary Standards Bodies
Academia, Coalitions, Partnerships
Suppliers, Customers, Competitors, Consumers
Media, Communities
Atmosphere, Hydrosphere, Biosphere

Stakeholder Groups

Stakeholder groups can be internal groups loosely defined as workers, contractors, or leadership, or narrowly defined as human resources department, health and safety professionals, for example. They can include committees, like the Health and Safety Committee or the Neuro-diverse Affinity Group, or Board Renumeration Committee, or Audit Committee.

Stakeholder Groups may also be external groups such as those listed in the call-out box: Financial Institutions, NGOs, or Communities. The authors have also seen many references on websites and sustainability reports to Earth, and the environment being a stakeholder due to the significant impact businesses have on the environment and the environment has on businesses – just watch your inbox next Earth Day!

Stakeholder Mapping

Taking from our business and marketing counterparts and applying to sustainability, stakeholder mapping is a formal assessment of stakeholder groups to understand material issues and drivers (Mural 2024). (For a more complex discussion on social sustainability and stakeholder engagement, visit accountability.org to review their AA1000SES global stakeholder engagement standard and supporting documentation (Accountability 2024).)

A well-developed stakeholder map informs the engagement strategy and embedded messaging. The communications strategy (a.k.a the engagement strategy) includes details on not only how often to communicate to the stakeholder and about what, but through what medium. The communication strategy is tailored to reflect not only the level of sophistication of the stakeholder group and the material issue, but the group's readiness to absorb content or the message your organization wants to convey. The messaging may also change over time, so there may be multiple champaigns (phases) in the communication strategy. The mapping exercise can be completed at the group or individual level. At a group level, the mapping will be helpful to understand similarities and differences in messaging to each stakeholder group. At an individual level, the stakeholder map may be used to identify circles of influence.

In this chapter, we will note that organizations, or you as the practitioner, may undertake stakeholder mapping at any point in the sustainability journey to as you are starting out, to reinvigorate the strategy and realign with stakeholders, or formalize processes – sometimes when programs begin to go awry, taking a small step back to formalize what has been done thus far can identify a small misstep thereby allowing for quick correction to get you back on track.

The Seven Steps to Stakeholder Mapping

1) Define the Purpose
2) Identify Material Issues (and Goals, if appropriate)
3) Identify Stakeholders

4) Determine Level of Engagement (High, Medium, Low)
5) Determine Level of Influence
   a) Primary, Secondary, Tertiary, Quaternary
6) Determine Stakeholder Engagement Category
7) Develop Engagement Plan
   a) See Communications Plan Sam

The Stakeholder Mapping exercise may be conducted for similar purposes as the materiality assessment: to inform strategy, inform disclosures, and/or to develop an engagement strategy. Depending on the purpose, different stakeholder groups may be selected in the assessment. Each stakeholder group is then qualitatively evaluated for their level of engagement across the identified material issues and, if appropriate to the exercise, for the goals they have externally communicated. A determination is made as to whether the stakeholder group is highly involved (or engaged), involved at the edges, or not particularly involved at all in the identified material issues. The stakeholder's level of influence is also evaluated. Consider for this section we are thinking about larger sustainability initiatives that extend beyond the organization's fence. Another consideration under the influence to reflect upon is how individual stakeholders may influence other stakeholders purposefully or tangentially in the market, for example, peers or competitors. And once all stakeholders are identified, racked, and stacked for engagement by material issue, you are ready to categorize the stakeholders and develop your engagement plan. See Figure 10.1 "Stakeholder mapping template" for an example template you can use to guide your discussions.

Each organization will be required to define each level of engagement (high, medium, low, and perhaps none) and level of influence (primary, secondary, tertiary, and quaternary) prior to initiating their mapping exercise. Primary stakeholders may be considered based on their level of influence as shown in this example; you may have also seen other models where the primary stakeholder is also defined by those having direct influence versus the secondary stakeholders that have only indirect influence and tertiary stakeholders with the least amount of influence. Quaternary stakeholders in that model have no influence at all, as we have also shown, however, are included due to their interest in the business impacts (Mural 2023).

If your organization does not routinely engage with external stakeholders, you may create an engagement rubric based upon publicly available information you are able to locate on your stakeholder. The goal is to create an objective scoring methodology that allows you to evaluate each stakeholder and how you may engage with that group in the future as you begin

Goal of stakeholder map: 2025 materiality and goal setting

Material issue: clean drinking water | Date evaluated: _3/5/25___

	Stakeholder group	Level of engagement	Level of influence	
1	City of Greenspace	3	5	15
2	George River District	5	3	15
3	TV 10 News	5	5	25
4	Rolling Hills Elem Sch	1	3	3
5	Rolling Hills Mutual Aid	3	9	27
6	Rolling Hills Fire Department	9	3	27
7	Engineering Department	3	3	9
8	Health, Safety, and HazMat Team	9	1	9
9	Environmental Stewardship Dept	9	5	40
10	Sustainability Committee	9	5	40
11	Executive Leadership Team	5	9	40
12	George River Environmental BoD	5	5	25

NOTES:

The George River Environmental BoD outgoing Chair is former colleagues with the LT at the Fire Dept where we have significant engagement through our confined space training program. Our communications and engagement strategy should use that "circle of influence" to share our message on clean drinking water and work around improving safe access to water.

Engagement, influence
1 = no contact, power to influence
3 = limited contact, limited influence
5 = sporadic contact, some influence
9 = regularly in contact, significant influence

Figure 10.1 Stakeholder mapping template.

to explore communication opportunities. Initial engagement strategies may well begin with monitoring stakeholder groups until sustainability initiatives and culture begin to take hold.

The Stakeholder Engagement

The structured process begins by identifying the stakeholder groups, qualitatively rating the group's interest and ability to influence. Once all stakeholders are mapped to their appropriate category, the modeler can look for opportunities to inform the communications strategy. Based on the level of interest and influence, we have four resulting categories monitor, show consideration, keep satisfied, and engage. See Figure 10.2 "Stakeholder engagement matrix" (Marquette and Windsor 2024).

- *Monitor* – low interest, low influence: this group of stakeholders has shown little interest in the material topic(s) and little ability to influence significant or strategic stakeholders. It is important to carefully document these qualitative decisions to ensure that all relevant relationships and material issues were considered and revisited over time. This may be particularly useful in industries with heavy merger and acquisition activities where relationships between organizations change.

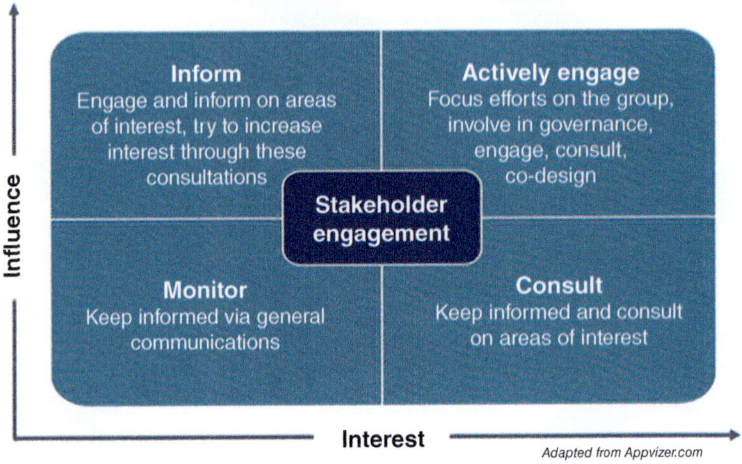

Figure 10.2 Stakeholder engagement matrix. Source: Marquette and Windsor (2024)/with permission of Woodard & Curran.

- *Show Consideration* – low influence, high interest: this stakeholder group has been actively engaged in the conversation and therefore may pick up on intelligence your organization might not otherwise have heard to be sure to show appropriate level of consideration; however, the group has a limited influence so their intelligence gathering capabilities may not yield results.
- *Keep Satisfied* – low interest, high influence: highly engaged stakeholders may well be at the right place when you need them, so keep them informed where appropriate, but do not place high expectations on their level of interest with your strategy as their level of interest may be elsewhere at the time.
- *Engage* – high interest, high influence: this highly interacted, engaged stakeholder group may well have significant influence both internally and externally. Keep them engaged and seek their feedback as this group may have circles of influence well beyond your organizational walls and may be able to share valuable feedback from groups that would otherwise not engage.

Each of the four stakeholder categories will have a unique engagement strategy overall; there will be similarities throughout each of the strategies at different junctures. For example, in the Stakeholder Mapping Template, we note that the Rolling Hills Fire Department is very engaged with our organization and has little influence over our strategy directly. However, due to connections within the Department and the local water District, the

Fire Department may be a strong voice on our behalf and use their "circle of influence."

Circles of Influence

While interviewing Tanis, she shared a lesson she learned early in her career that is applicable both to stakeholder engagement and really across sustainability. Circles of influence. Circles of influence are connections between individual stakeholders or stakeholder groups that can be influenced to modify the perspective of another individual or stakeholder group. Because the second stakeholder has greater trust with the first stakeholder, there is a greater potential for the second stakeholder to stay close to the first stakeholder and really listen to the ideas and introduce a few new ideas of their own. When the second stakeholder shares new ideas with the first stakeholder, the circles of influence spread each other becoming something better than themselves and yet still the same and aligned.

Circles of influence can be used within stakeholders and coalitions to build alignment and agreement to solve material issues. See Figure 10.3 "Stakeholder level of involvement" (Marquette and Windsor 2024).

Who Is Generally Involved?

Stakeholder mapping is often completed by the marketing team but has significant value for the sustainability and EHS teams. The mapping exercise can be completed easily and held in reserve as a planning tool versus a

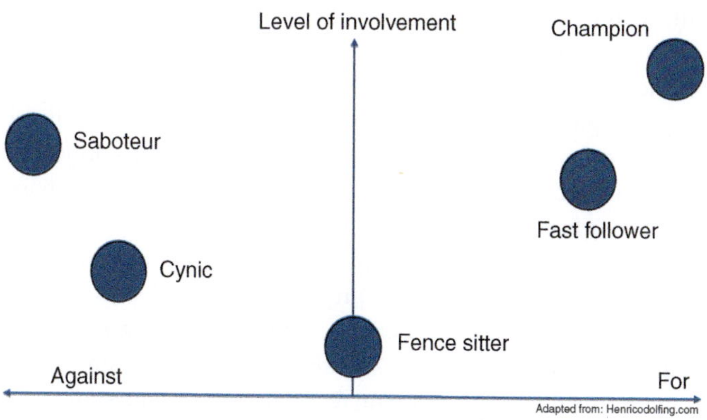

Figure 10.3 Stakeholder level of involvement. Source: Marquette and Windsor (2024)/with permission of Woodard & Curran.

document to be shared. However, when mapping larger stakeholder groups, it is helpful to talk through the evaluation to spur creative means to communicate with challenging stakeholder groups.

Where Is It Interconnected?

Workers are a specific stakeholder group. When considering influencing change and creating initiatives and opportunities, understanding the stakeholder group as a whole and specific influential individual within the group can increase your potential for engagement and success. Implementing sustainability within organizations is about culture change as previous chapters have covered. One aspect of culture change and stakeholder mapping is understanding the disposition of our stakeholders.

As health and safety leaders, we intuitively use stakeholder mapping and circles of influence as we engage with workers on health and safety issues. The most valued worker is the worker who is able to lead others. The leader may not be a worker with a significant role or responsibility but someone that nearly all workers listen to each day. If the leader is working safely, that worker influences the workers around them. Similarly, in the sustainability space. The leader may influence others to adopt responsible resource management. We call these leaders, champions.

As we embark on new endeavors, we need to be wary of a few other stakeholder personas. Refer again to Figure 10.3 "Stakeholder level of involvement" (Marquette and Windsor 2024) to review stakeholder personas.

Adapted from Henrico Dolfing, there are five stakeholder initiatives-dispositions: The Champion, the Fast Follower, the Fence Sitter, the Cynic, and Saboteur. Again, seasoned professionals have picked up on these personas lurking among us, but for our newer colleagues, we will acknowledge their existence in hopes we save them a little on the project learning time. Before you begin, try to identify those who may be, intentionally or not, the saboteurs or cynics. These personas can quickly derail a project or initiative beyond repair, but if you can win these personas over, you have found yourself a new fast follower, if not a new champion! How do you win them over? Often, it is through sharing, openness, and building trust. Give them the opportunity to understand the project, share the details in their terms, that is, meet them where they are and show them the value of the project through their eyes. Enlist the assistance of your HR team to give you support in finding the right words and techniques to ask questions and encourage communications. The authors will admit that it takes transparency and trust to win over some of these difficult personas and you would not be successful every time, but knowing to look is sometimes half the battle. We wish you every success. We need to acknowledge that saboteurs, cynics, and even fence-sitters exist, but we would not solve the challenge they pose here – this is a book about sustainability, after all.

What we will focus on, however, is leveraging the champions and fast followers to our advantage to get strategies and culture change, perhaps more quickly than it may have. Your goal early in any strategy development is to identify those stakeholders in the organization that are excited about the material issues and are striving to ensure your and the organization's success, they are your champions. This allows quick wins. Win: wins so to speak. Engage those individuals that are passionate, but perhaps are not in a sustainability or EHS role. Ask their feedback; get their insight. Next, the fast follower. Those who are anxious to get engaged, but need a little more direction, are the fast followers. The fast follower might shadow you or the champions to learn more about sustainability, EHS, or the strategy to be able to support the efforts. The fast follower will be a great help to safety or sustainability committee investigative efforts, for example. The fence sitters are just that. They will sit back and watch and wait, not engaging until they have a clearer understanding of whether success will be realized. They may be supportive, and then, again, they might not. As you are developing your circles of influence internally and asking for feedback and input, engage with these internal stakeholders, and deploy accordingly. You will be pleasantly surprised.

Why It Has Value?

The mapping forces the modeler to consider each individual stakeholder group and, in some cases, individuals within stakeholder groups. The evaluation, while qualitative, does provide an objective window into how to engage with various stakeholder groups and how to identify what their priorities are and why their priorities may not be aligned with the sustainability priorities.

The stakeholder map is the foundational document for the development of a comprehensive communications plan. The communications plan details with whom to communicate, how, and how often. As the sustainability strategy is adopted, the stakeholder influences and communications plans should also mature. See Figure 10.4 "Stakeholder influence on sustainability strategy" to understand how different stakeholders can influence the sustainability disclosure strategy.

As this graphic above demonstrates, organizations can be influenced by stakeholders across all aspects of their sustainability program implementation from policy, program and procedures development, initiatives identification and execution, disclosures and reporting, and investment and capital

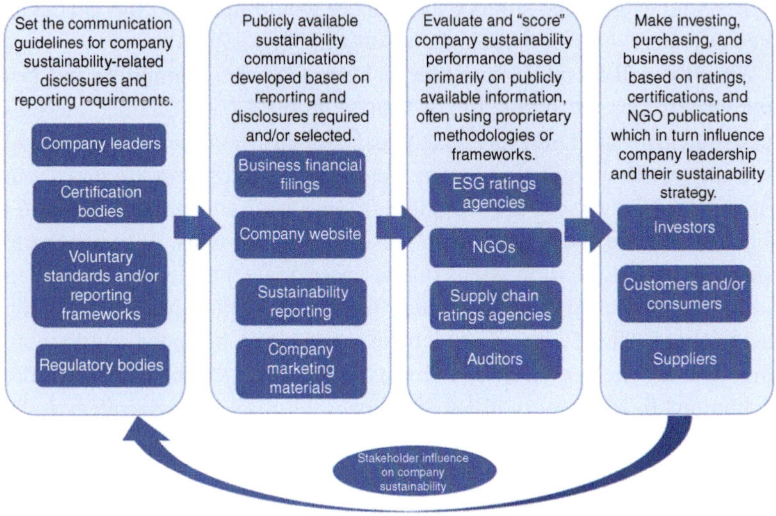

Figure 10.4 Stakeholder influence on sustainability strategy. Source: Marquette (2023)/with permission of American Society of Safety Professionals.

expenditures. Whether directly engaged with the organization or providing a public perspective on material issues, various stakeholders influence sustainability and present an opportunity for targeted communications.

And, to assist with communications planning . . . See Figure 10.5 "Communications Plan Sam."

We introduce Communications Plan Sam with special thanks to Jeanee Dudley and Taylor Roberge at Woodard & Curran. This fun play on a childhood memory should help take communications planning stick (hopefully without buzzing) in your mind!

1) Objective
2) Context
3) Target audience profile
4) Messengers
5) Mode of communication
6) Media-specific strategy
7) Message for each unique audience
8) Call-to-action
9) Cadence
10) Monitoring KPI

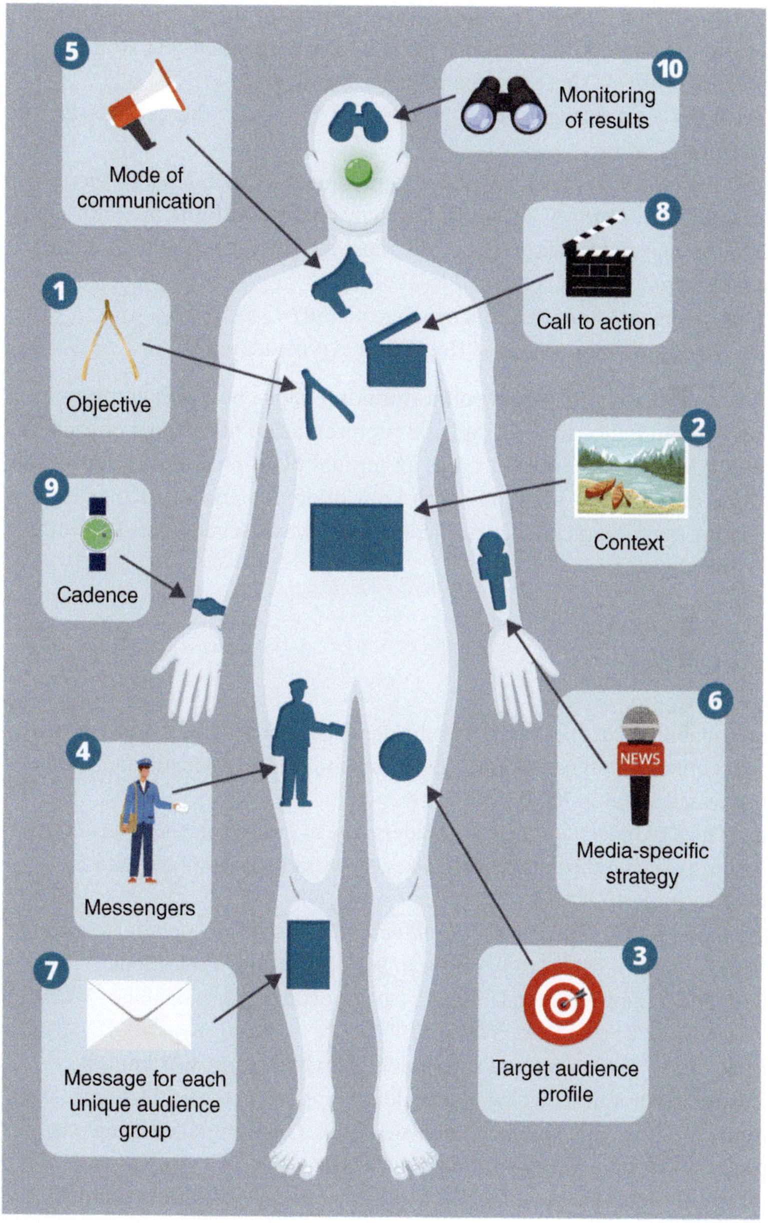

Figure 10.5 Communications plan sam. Source: Marquette and Windsor (2024)/ with permission of Woodard & Curran.

Objective – What is the overarching goal of the communications engagement?
Context – How does the objective pertain to the organization's strategy?
Audience Profile – There may be multiple profiles.
Messengers – Messengers may be engaged based on the profiles for the communication.
Mode – Email, letter post, TV advertisement, video, social media, etc.
Media-Specific Strategy – Does the strategy change with the medium?
Audience-Specific Strategy – Does the message change with the audience?
Call-to-Action – What is the call-to-action?
Cadence – How often does this message go out?
Monitoring – What is success? How often is it measured?

The schematic introduces you to thinking about how you communicate with your audience with a graphical representation of the goal of each element, which can be introduced as an annual plan, phased strategy, or any unique engagement effort to reach your audience and to further borrow from our marketing colleagues, ideally seven ways, seven times to reinforce your message(s).

References

Accountability (2024). AA1000 stakeholder engagement standard. https://www.accountability.org/standards/aa1000-stakeholder-engagement-standard/ (accessed 3 March 2024).

Cerri (2024). Understanding stakeholders: key to project success. https://cerri.com/understanding-stakeholders-key-to-project-success/ (accessed 12 February 2024).

Marquette, T. (2023). *Safety and Sustainability: Safety Professionals Handbook* (ed. J. Haight), 1171–1185. American Society of Safety Professionals.

Marquette, T. and Windsor, D. (2024). Operationalizing sustainability. Webinar, Woodard & Curran Host, 5 June 2024.

Mural (2023). How to create a stakeholder map [templates & examples]. https://www.mural.co/blog/stakeholder-mapping (accessed 03 March 2024).

Mural (2024). Stakeholder map template. https://www.mural.co/templates/stakeholder-map (accessed 03 March 2024).

Oxford Learners Dictionaries (2024). https://www.oxfordlearnersdictionaries.com/us/definition/english/stakeholder (accessed 30 June 2024).

11

Benchmarking

Introduction

As organizations embark on an effort to define their strategy, they often inform their development activities by benchmarking. While the Model described in this book (see Chapter 8 *Beginning the Journey*) shows concentric circles building upon each other, it is important to acknowledge many organizations have adopted their own unique methodology for embracing change and integrating new principles into their business' strategy. Therefore, it should be noted that benchmarking activities may occur within different elements of the Model. They may also occur across multiple elements of the Model. Benchmarking may occur at different times for different reasons by different teams. For our purposes, we will discuss benchmarking as a methodology in its whole without regard for when it is conducted or by whom.

In this book, we provide a step-by-step benchmark assessment processes; however, the authors want to acknowledge some benchmark studies are as simple as documenting in the margin what your peers are sharing externally, from metrics to the charitable giving amounts to any other aspect of sustainability your leadership is interested to learn more about and incorporate into your strategy (Robertson 2021). Those inputs may define your metrics, your charitable contributions, or future programs (Blackburn 2007). There is no requirement that benchmarking takes on a formal process.

Connecting the Dots: Environment, Health, Safety, and Sustainability,
First Edition. Kathy A. Seabrook and Tanis J. Marquette.

From ***Merriam-Webster***,

a benchmark (noun) is something that serves as a standard by which others may be measured or judged, a point of reference from which measurements are made, a standardized problem or test that serves as a basis for evaluation or comparison;

or (verb) to study (something, such as a competitor's product or business practices) in order to improve the performance of one's own company

Organizations may choose to compare their vision, goals, initiatives, or the level of effort against peers, competitors, or leaders in your sector or an organizationally defined set of organizations. The benchmark can confirm alignment with your peers or perhaps lead to the development of a strategy that will allow you to lead in the industry, providing competitive advantage. Benchmark assessments, or studies, may yield unexpected information – a peer organization may be doing far more work in an area than was previously identified or quantified (Robertson 2021). As you might imagine, given the diversity of topics within sustainability, the breadth of global industries, and how individual organizations, whether they be non-governmental agencies, companies, or federal or local governments, compare themselves to their identified peers, the possible content for a sustainability benchmark is endless. However, one thing remains. Ultimately, the benchmark defines a current state and opens the door to possibilities.

What Is It?

The benchmarking component of the Model element three "Sustainability Assessment and Implementation Tools" is a strategic comparative-analysis process with multiple decision points and variation in application. A benchmark assessment for sustainability typically identifies what needs to be understood or learned, sources of information (from interviews, literature search, to web-crawling), which peer group's performance or disclosures may provide insights, and definition of the measurement criteria to ensure objective assessment the organization against the criteria. The measurement criteria may also be known as assessment rubric, criteria rubric, or many other unique-to-organization descriptions. For ease of reading this chapter, we will use the term *scoring rubric* to describe "the defined measurement criteria" for evaluating information analyzed as part of the assessment.

In the authors' experience, most organizations use benchmark assessments to identify material issues, reporting, and disclosure topics to identify and define performance metrics, indicators, and goals. An organization may also use the benchmark results to identify collaboration partners (Sharma 2022).

Industry or trade association reports are a wealth of information and, depending on the scope of the assessment, may include valuable research information on peer companies. Sector associations like the American Chemical Council (ACC) or the Global Environmental Management Initiative (GEMI) provide opportunities for organizations to come together and share challenges in committees and work collaboratively toward solutions. The deliverable from committees, task groups, or working groups, may include benchmark assessments, or valuable resources for professionals outside the organization to access. The Global Environmental Management Initiative (www.gemi.org) working groups have developed tools to further environmental practices and sustainability initiatives since its inception in the 1990s. Similarly, the automotive industry has partnered together to benchmark their performance in the Suppliers Partnership for the Environment (www.supplierspartnership.org) and produced numerous resources including EV battery safety guidance documents that are available on their website. See Figure 11.1 for the complete Benchmark Assessment Process with each step and its outputs.

Identify What to Benchmark – Purpose and Scope

A benchmark assessment has a purpose and scope. The purpose is the question the benchmark will answer, for example, "How many of our peers have set Net Zero goals?" or "What aspects of ESG have our competitors implemented? The scope is the breadth or how wide to cast the net to get information and answer the question or address the purpose. Will the benchmark consider general sustainability topics, material topics, or other aspects of the sustainability strategy, like reporting or messaging. If the organization anticipates contracting a consulting or professional services firm to conduct the benchmark, the scope decisions may be finalized later in the process after the resource has been defined. As all organizations are unique, the actual order of each of these steps may be more refined in one step versus another; however, prior to hiring a third-party firm, the organization likely has defined at least a preliminary purpose and scope for the engagement (or project as it may be called). (At a minimum, the internal working team should be identified with the key decision makers and contacts noted

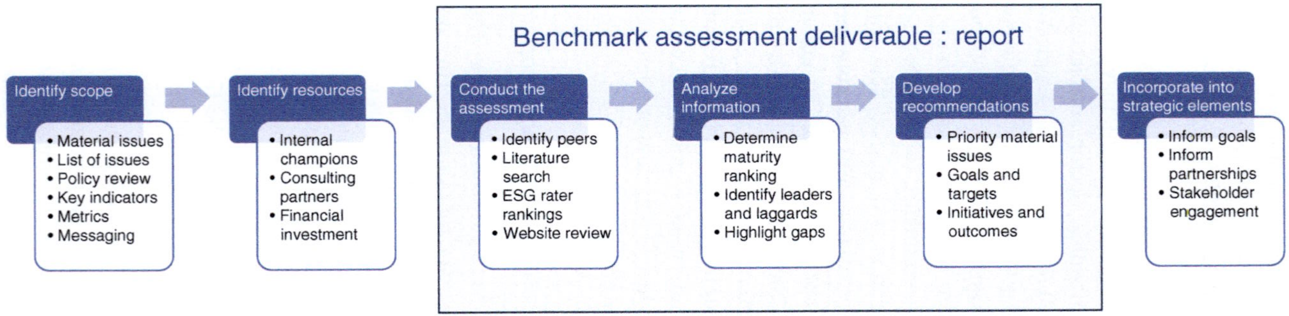

Figure 11.1 Benchmark assessment process.

prior to any outside engagement to ensure alignment and consistent communications.)

With acknowledgment that the scope may expand or retract as the benchmarking process continues, during initial considerations, the organization can self-define topics for inclusion in the benchmark, consider a third-party list of topics, such as the publicly released set of material issues, by industry sector, as defined by the Sustainability Accounting Standards Board (an entity within the International Sustainability Standards Board [ISSB]), or engage with stakeholders to create the list of topics of most interest to the stakeholder groups.

Similar to chemical companies reviewing the American Chemistry Council's (ACC) injury statistics to determine top quartile performance, sustainability benchmarks may consider similar quantitative performance data points, for greenhouse gases (GHG) or incorporate a much broader spectrum of quantitative and qualitative information like environmental stewardship goals and charitable giving or activities. The information needed to address the benchmark assessment purpose, whether narratives or performance metrics may be incorporated into organization decision-marking around goals and targets. It will provide insight as to what types of initiatives the industry peers are undertaking, including programs and policies that are underway or developed. Once the purpose is well understood, the organization will turn its attention to the scope or level of detail necessary to capture sufficient information from which to draw conclusions.

How does an organization decide on the breadth of the assessment? This too varies depending on the strategy the organization chooses to employ. A shorter list of topics may be considered if the benchmark is the first of its kind for the organization and the organization is immersed in multiple other new initiatives. It is speculated that by reducing the number of topics workers and stakeholders will have a greater willingness to participate and be encouraged to provide more feedback (Sharma 2022). The authors believe some of that willingness to engage is also based upon strong communications engagement and inclusion strategies. Organizations having already completed benchmark assessments or ones that are further in their sustainability journey, may choose to narrow the list to focus on a specific aspect of their strategy, or conversely, they may expand their benchmark scope as they now have champions and teams established across the organization that are ready to delve deeper into material issues. There is no one right approach other than the one that allows the organization to move forward. From the old adage, "*Rome wasn't built in a day*," for the professional, it is much more important to begin the benchmark process than be stuck in scope definition (Traverner 1190)! Take what you can get and move on.

Experience has shown the authors that you will no doubt have other opportunities to expand the discussion.

A benchmark may include a review of pertinent policy statements to inform how an industry sector or a peer company has defined a specific sustainability element. It isn't enough just to have a policy statement, details matter ... sometimes what is contained in the policy is the meter stick. Similar to a health and safety policy, stating worker safety is paramount with no other definition or clarification is likely insufficient. Sustainability policies may provide insight into how an organization defines sustainability. That is, does the organization strive to incorporate aspects of environmental protection and social responsibility into all business decisions, or, is the definition narrower and the organization emphasized environmental stewardship and compliance?

The benchmark may also evaluate the organization against a group of peer companies to determine leaders and laggards. The benchmark may consider organizations in your specific industry, leaders regardless of industry, or your specific value chain. Again, external raters and ranker publications may provide some insight, but typically, the organization will define for itself what constitutes a leader or laggard – or it will be an output of the benchmark. Across the value chain, an organization may choose to identify organizations that would benefit from collaboration or use the benchmark to identify organizations from which to learn. Benchmarks can also evaluate sustainability aspects from program implementation to goals to disclosures, but regardless of what the benchmark assesses, the assessment creates a current state view.

Peer Groups

Publicly traded companies are evaluated across numerous benchmarks by entities called ESG Raters, for example, State Street, MSCI, and Sustainalytics. MSCI's About MSCI ESG Research Products and Services section states they are "designed to provide in-depth research, ratings, and analysis of environmental, social, and governance-related business practices." ESG Raters develop their list of peer companies based upon companies of similar revenues and market cap facing similar "key issues," not necessarily always by product or industry, although each rater has their own proprietary definition. MSCI's ESG Ratings Guide for Issuers states that it has greater weight in the MSCI peer company groupings (Investopedia 2024a;

MSCI 2022). When reviewing ESG Raters reports, an organization may find some consistency in the list of companies identified as peer companies, however, from report to report the list may change. The specific ESG Rater the organization parallels may have been identified based on the rater-of-choice of the primary investor(s). Senior company executives also have a consistent set of companies they use as their benchmark for financial performance (Investopedia 2024b). These different lists of peer companies, aside from the C-Suite's list, which may not be public, can be found in public ESG Raters website searches or by searching a company's 10-k United States Securities and Exchange Commission (SEC) filing. For consistency and familiarity, sustainability professionals may choose to use the same set of companies in their benchmark so leadership can compare sustainability benchmarks to other business information they are using.

Resourcing the Benchmark

Benchmark assessments can be completed using publicly available information only, a combination of public information and direct information, or by directly contacting benchmark organizations.

Organizations may choose to develop their benchmark assessment in-house using publicly available information, trade association information, or third-party rating reports, for example, MSCI or paid market research reports. In-house assessments may be chosen to reduce financial costs, ensure internal data privacy, to reduce time commitments, or for any number of unique-to-organization reasons. Regardless, once an in-house effort has been selected, the assessment team will need to be identified. In-house teams generally follow established internal project management methodologies to implement benchmark assessments.

The organization may also work with a consultant to develop the benchmark view. Consulting practices and professional services can include conducting the benchmark assessment with deliverables that vary nearly as much as sustainability has topic areas. Many consultants use proprietary methodologies, for example, S&P Global's Benchmarking Solutions will "support you with visualized performance trends, actionable data, and gap analysis, and direct access to peer practice examples." Small boutique professional services offer a more intimate engagement, and like the larger firms, come with a range of sustainability experiences. The opportunity to partner with a larger consulting firm with access to sustainability ranking platforms may be an option for some organizations with greater funding for this endeavor. Many organization-specific reasons factor into the final selcction of a consulting partner, but regardless of whether your organization partners with a boutique consulting firm or one of the Big 4, consultants will follow defined methodologies and provide a robust output report. Advantages of working with a consulting partner include, but are not limited to, neutrality, expertise, and predictable costs (Green Business Bureau 2024). Consultants are not entwined in the company politic, usually have multiple experts that can support your effort, and, once the contract is signed, the investment and the timeline is set. A consultant partner may also be able to gain more insightful voice of customer information than the assessing organization due to perceived differences in confidentiality. Understanding the scope and goal of the assessment, as well as available time and resources, will help you decide what approach is best for your benchmark assessment and who should be involved with your consulting partner.

Conduct the Benchmark – Competitive Analysis

Beyond topic areas, sustainability benchmark assessments may consider goals, policy statements, or even implementation maturity across an industry. For example, if the organization was striving to understand the amount, type, and depth of health and safety training provided to a workforce, comparing it to organizations with a similar work environment may be appropriate. An apples-to-apples comparison. However, if the organization was striving to understand what a leading organization reports for their annual disclosures, the organization may select companies from various sectors and sizes that have been identified as leaders, by others, which may be a peer or by raters and rankers – the entities that publish annual lists of top performers, based on their unique criteria. The benchmark purpose and scope can also assist decision-makers in the formation of the comparative aspect of analysis. Organizations may also choose to include multiple peer groups in their assessment. If, for example, the purpose of the benchmark assessment was to inform sustainability messaging, the message to customers may be different from suppliers; to adequately benchmark the appropriate messaging, the peer groups' perspectives may be isolated to ensure messaging is appropriate for each group. See Figure 11.3 for Benchmark Competitive Analysis Tool excerpt.

Conduct the Benchmark – Information Gathering

The benchmark may be conducted as a literature review. Each organization in the peer group is evaluated against the set of criteria as identified by the benchmark purpose and scope. The literature review sources vary by assessment and may include company websites, documentation on those websites, or by creatively searching the interest.

The assessing organization may also determine their information is best gathered through a questionnaire. The questionnaire length is purpose and scope dependent and very often also stakeholder dependent. Questionnaires, like literature searches, have benefits and challenges (PointerPro 2023).

Each question on the form has an aligned scoring which will culminate in an objective overall score for the responding organization. The questionnaires are generally distributed via email for ease of implementation, either with a link to a web form or a fillable PDF that can be completed and returned. However, sourcing the email distribution list may also pose challenges and may also be impacted by privacy laws.

Another way to gather benchmark information is through direct contact of interviews. Interviews conducted with actual customers are also known as the voice of customer (VOC). For the purposes of sustainability, the VOC definition is expanded to include stakeholders, of which some may be customers. VOC exercises can be conducted in two steps with the first also being the questionnaire sent via email. Then, the second step is interviewing a select group of representative respondents to learn more about their responses. By adding the VOC interview step, the assessing organization can confirm patterns and trends visible in the raw data.

VOC interviews are typically very insightful (there is a dialogue rather than a unidirectional response and questions can be answered in real time); however, they may require increased diligence from the interviewer to ensure objective output. Companies that have chosen to conduct their assessment in-house, may hire a consulting company to conduct the interviews as a way to address confidentiality challenges as noted in the Pointerpro list of pros and cons. See Figure 11.2 for pros and cons of questionnaires.

Process – Analyzing Information

The benchmark may only ask if the peer organization has a policy, or the benchmark may consider more specific details and applicability of thepolicy.

PROS	
Questionnaires are versatile and efficient instrument for data collection	
Advantage	Explanation of Benefit
Large-scale data collection	A lot of information from a lot of stakeholders
Standardized response	Predetermined answer options
Anonymity	Unbiased answers without fear of repercussions
Cost-effective	Large pool of information with low cost
Data quantification	Responses converted to numbers to identify patterns or trends
Flexibility	Customizable content for specialized information needs, respondent diversity, or data accuracy
Quantitative and qualitative	Simultaneous data collection provides for more holistic stakeholder insights
Ease of analysis	Ability to use statistical software to analyze data at speed
Assessment standardization	Standardization between assessments increases reliability and validity of the findings
Accessibility	Enables individuals with differing abilities, languages, or access participate, enhancing inclusivity
Time efficiencies	Substantial amount of information in a reasonable timeframe, ensuring time interpretation of results
CONS	
Limited in depth	Inability to capture nuanced insights due to limited depth of information
Response bias	Participant interpretation of questions may bias results
Low response rates	Lower response rates can bias the results of those that do respond
Misinterpretation	Participant may misinterpret questions leading to inaccurate results
Inability to capture non-verbal cues	Facial expressions and body language play a role in communication, questionnaires lack of non-verbal cues
Limited engagement	Participant's lack of interest, attention, or active involvement during completion process

Figure 11.2 Advantages and disadvantages of questionnaires. Source: Authors rendition, adapted from Pointerpro.

Lack of context	Participants may make assumptions leading to compromised data quality
Difficulty capturing complexity	Structured questions may not allow for full expression of thoughts, opinions or experiences
Inability to ask clarifying questions	Inability for assessing company to ask real-time follow-up questions to clarify participant's response
Sampling limitations	Participant sample group may not be truly representative of larger group
Measurement errors	Deviation between the true value of the concept measured and the participants response
Response fatigue	Lengthy, repetitive, overly complex questionnaires overwhelm participants and negatively affect data quality

Figure 11.2 (Continued)

The benchmark may note that the policy is referenced on the website, but not provided, or the policy may actually be available on the website, but limited in scope. Each element of disclosure should be clearly defined in the objective benchmark to enable consistent scoring of each organization.

To ensure the benchmark activities are objective and criteria is consistently applied, most benchmarks use external references to set at least part of the criteria; however, there is no requirement that a sustainability benchmark follows a specific standard, framework, or reference. To ensure each element is evaluated consistently and objectively, the best practice is to define success criteria in advance.

For example, a literature search documenting what sustainability information is shared on an organization's website could be rank defined, with five being the highest score, by whether a topic is:

1) Mentioned on the website,
2) Implied as having been implemented, based on other references,
3) Partially discussed on the website, but without significant details,
4) Discussed in detail, but does not meet all elements of a defined best practice, or
5) Covered in great detail, meeting or exceeding the best practice, with or without attachment.

See Figure 11.3 for mock webpage.

In the mock webpage example shown in Figure 11.3, the greenhouse gas emissions section would likely earn not greater than a two in our scoring

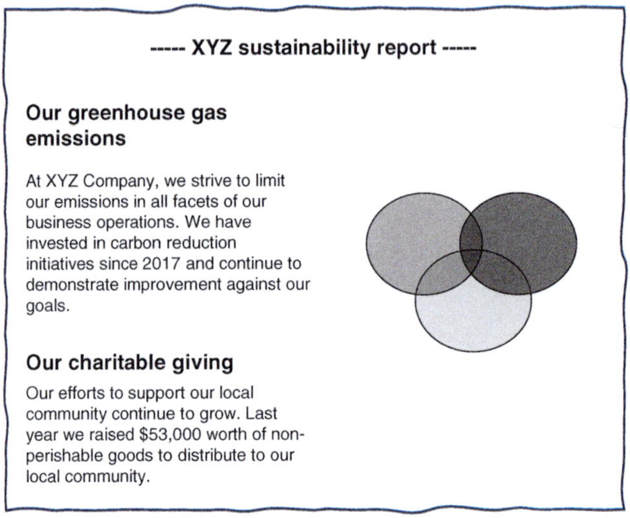

Figure 11.3 Mock website of sustainability disclosures.

rubric of our ESG Benchmark; however, the paragraph on charitable giving, while providing minimal details, those details are quantitative and thus could score a three or four on our rubric. The ESG Benchmark rubric serves as a guide to ensure consistent and objective scoring.

See Figure 11.4 ESG benchmark.

Analyzing the Data

The process of reviewing websites, white papers, ESG Rater reports, and, in some cases, internal policies programs, and practices will allow the assessment team to determine the maturity of the peer companies, identify leaders and laggards based on the pre-defined criteria, and highlight any gaps. Each assessment team, whether in-house or via a consulting firm, will have their own proprietary system and scoring matrices; the underlying premise is the same. See the example scoring rubric in Figure 11.4 "Example benchmarking template for scoring environmental stewardship elements" to visualize what the data analytics could look like. Each entity your organization has identified in its peer group, whether investors, competitors, customers, or recognized industry leaders, is scored across the scope of sustainability

ESG benchmarking template

Topic	Metrics	Company score	Company investors		Competitor scores				Customer scores				
			Investor A	Investor B	Competitor A	Competitor B	Competitor C	Competitor D	Customer 1	Customer 2	Customer 3	Customer 4	Customer 5
Enterprise level environmental management	Is there an enterprise level environmental policy?	4	3	5	3	4	5	5	5	3	3	3	5
	Is there a sustainability policy?	4	3	5	3	2	3	3	2	3	3	3	5
	Is there an environmental management system and are any policies disclosed?	4	3	3	3	0	3	0	3	0	0	2	3

Figure 11.4 Example benchmarking template for scoring environmental stewardship elements.

issues or impact areas. That scoring allows the analyst to identify the material issues, perhaps the goals of leading organizations, or a list of initiatives being implemented across the peer group. It is likely the information, if using questionnaires to gather data, will also lead to visible patterns and trends.

See Figure 11.4 “Example benchmarking template for scoring environmental stewardship elements.”

In the data collection process steps, the assessment team will have collected example documentation from the peer companies in review. These narrative reports are also evaluated and pertinent examples collated.

Develop Benchmark Report

Benchmark reports may or may not have recommendations depending on how the initial scope was defined; however, it will conclude with an executive summary and list of findings and analysis of the peers assessed. Ideally, the report will recap the process, documents, and data reviewed and provide a list of material issues and the process forward, if not recommendations. The report should serve as a roadmap for developing the list of material issues, stakeholder mapping (at least the initial inputs, and potentially draft strategies for consideration. The report should inform goals and targets, potential partnerships, and lay the foundation for stakeholder engagement.

In the example used in Figure 11.4, the Benchmark report would include the summary scoring, rubric scoring basis, and results by category in addition to elements discussed above. In the author’s experience, nearly all benchmark reports have some element of proprietary quantitative scoring embedded in the reports.

Incorporation into Strategic Sustainability

A benchmark report with a series of findings may be provided to an internal team to review and develop recommendations, or perhaps, the value of hiring the consulting firm was to gain from their insights and expertise through a robust list of recommendations.

Once the Benchmark Report is complete, organizations may continue partnership collaborations and develop actionable recommendations from the Benchmark Report. The organizations will further internalize the report and determine within a smaller leadership group how the report will be used. Ideally, the Benchmark Report informs the sustainability strategy in whole or part.

Organizations may be formal in their approach to the next steps or more fluid. The recommendations may include moving to the more formal materiality assessment, initiatives announced whose goal is to respond to gaps identified in the benchmark, or be directly incorporated into the written strategy. As will be discussed in Chapter 14, greenwashing has become a very visible conversation in the sustainability space; as such, documentation of decision-making processes and background rationale can become important substantiation stemming from the benchmarking assessment should your organization ever be subject to a greenwashing investigation.

Who Is Generally Involved?

There is no "right and only" way to conduct a benchmark assessment – much of the "who" will be determined by how an organization is structured, who the sustainability champions are or to whom they report, or at times, which department is funding the effort and the "what" of the assessment. Essentially, the benchmark methodology may be shaped by the subject, champion, and organizational resources, both financial and people. And, as we have discussed in this chapter, the who may well be determined by whether the assessment is conducted in-house or through a consulting firm.

An organization may also have been on their sustainability journey for some time but perhaps not have fully addressed all aspects of supply chain sustainability, for example. The initial foray into understanding supply chain impacts may accomplished through the benchmark. In this case, the team engaged in the benchmark could be very different from those engaged in understanding the breadth of peer company sustainability policies. The extent to which team members are involved in a benchmark effort may also be determined by the execution of the study. If the assessment is a literature review only, the breadth of individuals involved may be more limited versus a benchmark assessment executed by a questionnaire. In both these examples, the team could be an external third party or an in-house group.

If completing the benchmark questionnaire in-house, the sustainability team (whether that team is a committee, the chief sustainability officer [CSO], or other cross-functional group) should seek input from the legal and the commercial team to support the assessment effort. The legal department will assist with confidentiality matters and ensure compliance with international privacy laws. The chief counsel may also have input on the scope or documentation requirements for the assessment. (Depending on the scope of the assessment, for example, policy statements, respondents may be requested to submit documentation in support of their answers.) If the benchmark methodology includes a questionnaire completion, email addresses of customers and suppliers may be secured through the commercial, procurement, enterprise risk teams, and sent via the enterprise-wide software capability.

Where Is It Interconnected?

The benchmark assessment is interconnected across an organization in two ways. One is because benchmarks are used in other areas of business process and because the sustainability benchmark report, the output in this Model, may be used to inform decision-making across multiple functions in the organization. A benchmark may also use customer and supply chain information often found in the procurement, logistics, or commercial files or enterprise management software and its results used to augment their business benchmarking process. The environment health and safety team benefits from the operational performance metrics and data but also gains value through understanding the broader aspects like training, violations, and initiatives of peer organizations. In fact, health and safety performance may be material to the benchmark. (See Chapter 12 for more information on materiality.)

Why It Has Value?

Conducting a benchmark assessment may occur at any point in sustainability implementation, as we have noted. The initial benchmarking may not be a formal process but a reflection of what senior organizational leaders are "seeing" in the market. Your initial foray may be an opportunity to embed sustainable thinking into existing tools, methodologies, or programs. As programs evolve, a more formal benchmark may be conducted to confirm assumptions previously used to define the initial program elements. Even organizations that began with the formal benchmarking process may periodically reassess the current state of their internal programs or the industry to re-inform their strategic direction. The benchmark report becomes a record of the moment-in-time goals, policies, and material issues for the company; even the benchmark purpose and scope provide a view into how the organization was considering sustainability at the time. As the strategy and sustainability journey evolves over time, the benchmark reports laid side-by-side may provide insight into future trends in material issues (Freiberg et al. 2020).

The benchmark identifies initial steps, confirms current state, and provides future direction. If particularly challenging to garner support to conduct a benchmark assessment, the professional may also consider leveraging established business processes and analyses already in use within the organization. The author has modified the commonly used SWOT Analysis so named for its four quadrants of information: Strengths, Weaknesses, Opportunities, and Threats) for use as a sustainability benchmark tool. Using one quadrant as an example, the team assesses its strengths not in the market as the traditional SWOT may consider, but as pertains to environmental stewardship, social responsibility, and governance (ESG). While perhaps not as robust an assessment as the benchmark, directionally, the organization will have a place from which to grow.

Similar tools or methodologies that can also be customized to evaluate sustainability factors include:

- PEST Analysis (Political, Economic, Social, and Technological), which is "a management method whereby an organization can assess major external factors that influence its operation in order to become more competitive in the market" (Investopedia 2024c). The PEST analysis also has numerous iterations incorporating legal, ethics, and environment considerations: PESTLE, PESTEL, STEEPLE (Pestle Analysis 2024).
- SOAR, a play on SWOT, the acronym stands for Strengths, Opportunities, Aspirations, and Results, and (Digital Leadership 2024).
- SVOR for Strengths, Vulnerabilities, Opportunities, and Risks which incorporates more data and interconnectivity considerations than SWOT or SOAR.

When modifying tools in existence, recognizing the tool limitations is important. The SWOT and SOAR analyses do not identify the necessary underlying data; thus, if choosing either of these tools, the professional will need to consider how to incorporate performance data (either as an input or output) outside of those models.

Look across your organization and identify other evaluation processes embedded in different teams and functions and consider ways to infuse sustainability perspective in the outcomes. This enables sustainability implementation to become a value and core cultural component rather than a bolt-on-activity or initiative. Then, always take the opportunity to share back to the strategic team their tools with the sustainability information incorporated. The authors have found success building from what people know to bring something new, creating buy-in and engagement throughout the process not only with the cross-functional teams that participated in the process but with leadership as well.

Integration into Other Strategic Elements (Chapters)

The Benchmark Assessment report may become the foundation for the organization's sustainability programs, goal setting, or disclosure strategy. The duration for which an assessment report remains relevant will be organization and industry specific; it may be driven by internal activities or externalities. In this chapter, we have explored the Sustainability Benchmark report and alternate methods to arrive at similar outcomes. From decades of combined organizational and sustainability program implementation experience, the authors encourage the reader to always take the wins as you can get them and build toward your goals. Culture change, like Rome, does not happen overnight or because the CEO said so, change happens because champions, like you, can see the future state and have the tenacity to keep filling in the colors day by day, year by year.

References

Blackburn, W.R. (2007). *The Sustainability Handbook: The Complete Management Guide to Achieving Social, Economic, and Environmental Responsibility*. Earthscan.

Digital Leadership (2024). SOAR analysis template step-by-step guide for strategic planning. https://digitalleadership.com/blog/soar-analysis/ (accessed 28 January 2024).

Freiberg, D., Rogers, J., and Serafeim, G. (2020). How ESG Issues Become Financially Material to Corporations and Their Investors (November 4, 2020). Harvard Business School Accounting & Management Unit Working Paper No. 20-056, Available at SSRN: https://ssrn.com/abstract=3482546 or http://dx.doi.org/10.2139/ssrn.3482546

Green Business Bureau (2024). Is sustainability consulting better than an in-house team? https://greenbusinessbureau.com/blog/is-sustainability-consulting-better-than-an-in-house-team/ (accessed 27 June 2024).

Harvard Business School (2020). How ESG issues become financially material to corporations and their nvestors. Harvard Business School working paper 20-056.

Investopedia (2024a) What is the difference between market capitalization and revenue? https://www.investopedia.com/ask/answers/122414/

what-difference-between-market-capitalization-and-revenue.asp (accessed 30 June 2024).

Investopedia (2024b). Peer group: how it is used example pros & cons peer group: definition, how it's used, example, pros & cons (accessed 28 January 2024).

Investopedia (2024c). https://www.investopedia.com/terms/p/pest-analysis.asp (accessed 28 January 2024).

Merriam-Webster (2024). Merriam-Webster dictionary definition benchmark. https://www.merriam-webster.com/dictionary/benchmark (accessed 30 June 2024).

MSCI (2022). MSCI ESG ratings guide for Issuers.pdf (accessed 30 June 2024).

Pestle Analysis (2024). https://pestleanalysis.com/what-is-pestle-analysis (accessed 28 January 2024).

Pointerpro (2023). Questionnaire pros and cons. https://pointerpro.com/blog/questionnaire-pros-and-cons/ (accessed 30 January 2024).

Robertson, M. (2021). *Sustainability Principles and Practice*, 3rde. Earthscan from Routledge.

Sharma, H. (2022). How short or long should be a questionnaire for any research? Researchers dilemma in deciding the appropriate questionnaire length. *Saudi Journal of Anaesthesia*, 16(1): 65-68 https://doi.org/10.4103/sja.sja_163_21. Epub 2022 Jan4. PMID: 35261591.

Traverner, R. (1190). French collection of poems entitled "Li Proverbe au Vilain" 1190, first English translation accredited to Richard Traverner's 1545 Latin translation.

12

Materiality

Introduction

Materiality can be simplified into "what is important" from a sustainability sense and yet it is so much more. It describes a thing and a process. It is in this difference that the confusion begins. In this chapter, we will attempt to unpack that confusion and bring some clarity to the concept of materiality.

As it relates to this book, materiality is simply what is important to the company and its stakeholders from a sustainability perspective. Materiality, however, once determined (following the materiality assessment process), may be used to provide direction to an organization's overall sustainability strategy, key initiatives, and/or its approach to its disclosures. As sustainability and, more specifically, ESG is more intertwined with financial reporting, understanding the nuances of materiality as viewed from the financial sector becomes more critical (Reuters 2022).

How a specific sustainability issue is determined to be important and to whom, is through a materiality assessment. The materiality process seeks to identify sustainability issues that are important to specific stakeholders. These stakeholders may include local community members, customers, investors, academia, regulators, industry partners and competitors, workers, and consumers. In the authors' experience, which stakeholders an organization chooses to engage is often a reflection of both the stakeholders' sustainability maturity and the purpose of the materiality assessment. When the stakeholders are investors, the output of the materiality process may have financial and legal implications (Katz and McIntosh 2021).

For an investor, a material issue may impact the investor's decision to invest in an organization. More specifically, materiality was defined in 1933, by the US Securities Act as "those matters as to which an average prudent

Connecting the Dots: Environment, Health, Safety, and Sustainability,
First Edition. Kathy A. Seabrook and Tanis J. Marquette.

investor ought reasonably to be informed before purchasing the security registered." In 1976, the US Supreme Court ruled that an issue is material if there is "a substantial likelihood that a reasonable shareholder would consider it important in deciding how to vote," or "a substantial likelihood that the disclosure of the omitted fact would have been viewed by the reasonable investor as having significantly altered the 'total mix' of information made available" (Katz and McIntosh 2021; US Supreme Court 1976). It was further refined in 1999 to state "(a) matter is 'material' if there is a substantial likelihood that a reasonable person would consider it important" (US Supreme Court 1976).

Across materiality definitions, which continue to evolve, the financial-based definition has become known as single materiality. Single materiality focuses on the sustainability issues that impact the organization's ability to drive economic value for investors and shareholders. Rooted in financial and investor circles demand for greater visibility into sustainability issues, the original Sustainability Accounting Standards Board (SASB) materiality definition adopted with the US SEC (SASB 2016). As SASB merged with the Integrated Reporting framework and was ultimately absorbed by ISSB and IFRS, the alignment with the financially focused single materiality remained.

However, in its efforts to demonstrate willingness to recognize value beyond single materiality, in mid-2022, the International Financial Reporting Standard (IFRS) and GRI, formerly the Global Reporting Initiative, signed a "collaboration agreement." The agreement created the navigation document between their standards to enable users to align investor and multi-stakeholder-focused disclosures (GRI 2022). Their commitment was reaffirmed in May 2024 as the two frameworks announced their joint mapping and analysis for standards interoperability tools for greenhouse gas emissions and their sustainability innovation lab based in Singapore (3BLMedia 2024).

The GRI outlines requirements for disclosing material sustainability issues following their guidance documents related to impacts and multiple stakeholders that ultimately became known as double materiality. Double materiality combines financial materiality with externalities (impacts coming into the organization) to form the sustainability footprint of the organization. In 2020, GRI revised their definition for simplicity to "reporting on those topics that reflect its most significant impacts on the economy, environment, and people, including impacts on human rights"; however, users of GRI Standards are encouraged to apply double-materiality methodology in practice (GRI 2021).

See Figure 12.1 "Materiality definitions (Marquette 2024)."

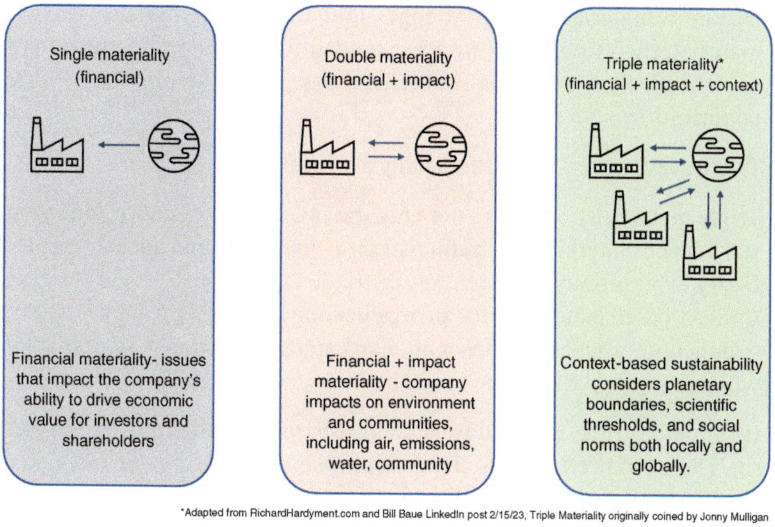

Figure 12.1 Definition of materiality.

Through two decades of implementing sustainability programs and attending sustainability conferences around the world, there are a few definitions of materiality that have risen to global discussion in the authors experience.

Types of Materiality

Through two decades of implementing sustainability programs and attending sustainability conferences around the world, there are a few different terms used to describe materiality that have risen to the global discussion.

Materiality became a little more confusing with the introduction of absolute and relative materiality context-based materiality and dynamic materiality (McElroy 2019). However, consider the core tenets of materiality and each become a little clearer.

- Absolute materiality guides an organization to consider whether the issue is material at all.
- Relative materiality prioritizes importance through a weighting and progression process.

- Dynamic materiality acknowledges that time may influence what is material; therefore, being strategic in identifying material issues is key to effectively managing material risk and identifying future opportunities.

This leaves context-based materiality and triple materiality:

- Triple materiality considers not only the financial materiality and material impacts, but the organization-specific thresholds and allocations necessary to be sustainable (McElroy 2019).
- Context-based materiality for an organization is, as explained by McElroy, "a function of what its impact on vital capitals are relative to norms for what they would have to be to ensure human well-being."

Essentially, triple materiality is context-based materiality in a simpler form. The referenced capitals align with the six capitals from the integrated reporting framework: financial, manufactured, intellectual, human, natural, and social. The use of the capitals within the context-based materiality assessment allows the organization to evaluate materiality in a manner that supports an integrated triple bottom line performance accounting (see Chapter 6 for more information on the capitals).

Work continues on context-based materiality through both Dr. McElroy's Center for Sustainable Organizations nonprofit, which can be found at https://www.sustainableorganizations.org/about-cso/ and his collaboration with r3.0, which can be found at https://www.r3-0.org/. The authors encourage our readers to explore both centers to learn more about what our colleagues are doing to advance materiality across sectors and organizations. Each of these advanced approaches to materiality, in the authors' opinion, considers the urgency necessary to address climate change by providing tools and resources to assist organizations in their sustainability journey.

One opportunity of note, recognized in the "concluding remarks" of Dr. McElroy's Making Materiality Determinations: A Context-Based Approach working paper for further exploration, is the context-based materiality assessment represents one organization. Here, McElroy acknowledges the materiality thresholds and allocations for just one organization, and the authors see that opportunity as another dimension for consideration – a materiality where the equation considers the denominators of the value chain perhaps, to truly understand the real impacts, the real sustainability indicators and performance.

The concept of mutual materiality is built on context-based materiality to include the interconnectivity between organizations and interconnected supply webs. The organization is not just acknowledging the planetary boundary, thresholds, and social norms between the company and the planet and one supply chain, but the interconnectedness between planetary boundaries, multiple supply chains, sectors, and the same company. The company may make products that transcend multiple supply webs that crisscross as they move through the industrial markets. Further, impacts may be more than additive, the author surmises, in some cases, they may indeed be exponential as we see in the health effects in the industrial hygienist's arena with the synergistic effect of some chemical exposures.

For sustainable success, the author argues that organizations must move beyond individual materiality silos and recognize the interconnected web of influences they share with industry thought leaders, partners, value chains, and stakeholder communities. Our reliance on planetary thresholds and allocations, as our colleague Mark McElroy has set out in context-based materiality, is beyond important. The United Nations (UN), through the proliferation of the Sustainable Development Goals, also recognizes the value of interconnectedness in their seventeenth goal – Partnerships for the Goals. Without recognition of a company's impact on others, whether it be locally or around the world, human well-being is not attainable or sustainable.

See Figure 12.2 "Mutual materiality (Marquette 2024)."

To explain, in Figure 12.2, for simplicity, we have one organization, Chemical Manufacture A supplying three different industries: silica-based consumer products, plastic-based products (industrial and consumer), and

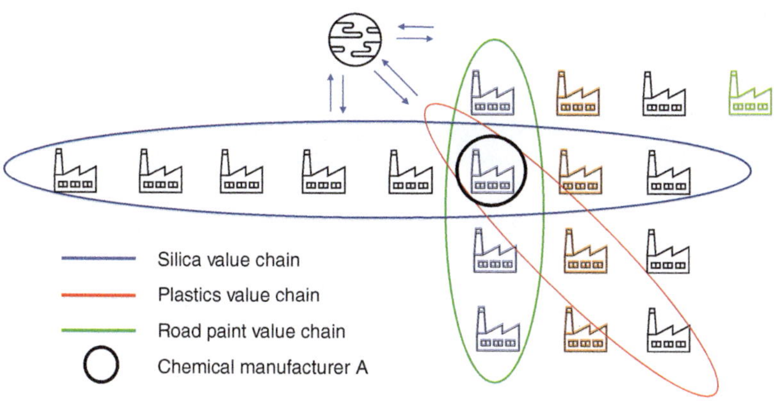

Figure 12.2 Mutual materiality.

road paint customers. Each supply chain has its own unique material sustainability issues and impacts, but also some potential overlap given they each purchase at least one raw material from the same chemical manufacturer. How these organizations collaborate, build capacity (educational and awareness), and come together to solve challenges may be an example of how they meet their internal and external goals and targets whilst recognizing jointly held thresholds and shared allocations through their ongoing stakeholder engagement. As authors, we look forward to the practice of *mutual* materiality coming together.

See Figure 12.3 "The progression of materiality (Marquette 2024)."

To test how we are stronger together when we share a mutual materiality mindset, the authors conducted a small experiment at the National Safety Council's Future of EHS 2024 Pre-Conference Workshop. First, the room was divided by generations and asked to rank five sustainability issues in order of priority; then, the room was divided by profession and asked to rank the same five sustainability issues while fully recognizing this is not a scientific study. It does, however, open the window for future exploration. When split by generation, the group ranked the material issues with some alignment; however, when the same professionals are grouped by profession (a more mutual perspective), the alignment between the groups is even more apparent. More investigation beyond this demonstration is necessary to confirm if the issues are truly mutually material, but the interconnectedness of the mutual professions versus the generations did show a change in the perspective of the same material issues with the same individuals.

See Figure 12.4 "The generational ranking within the mutual materiality experiment."

See Figure 12.5 "The profession ranking within the mutual materiality experiment."

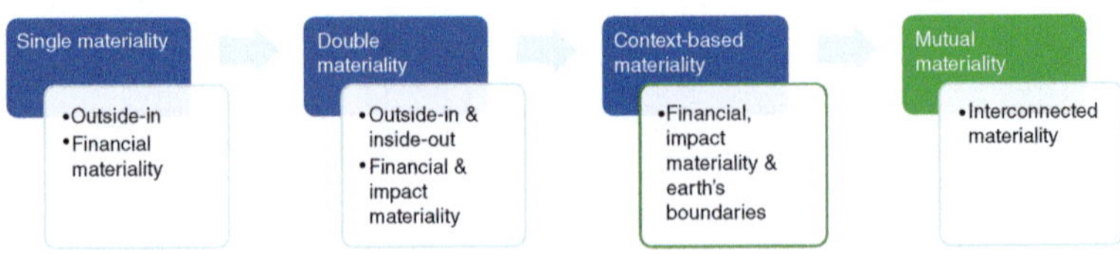

Figure 12.3 The progression of materiality.

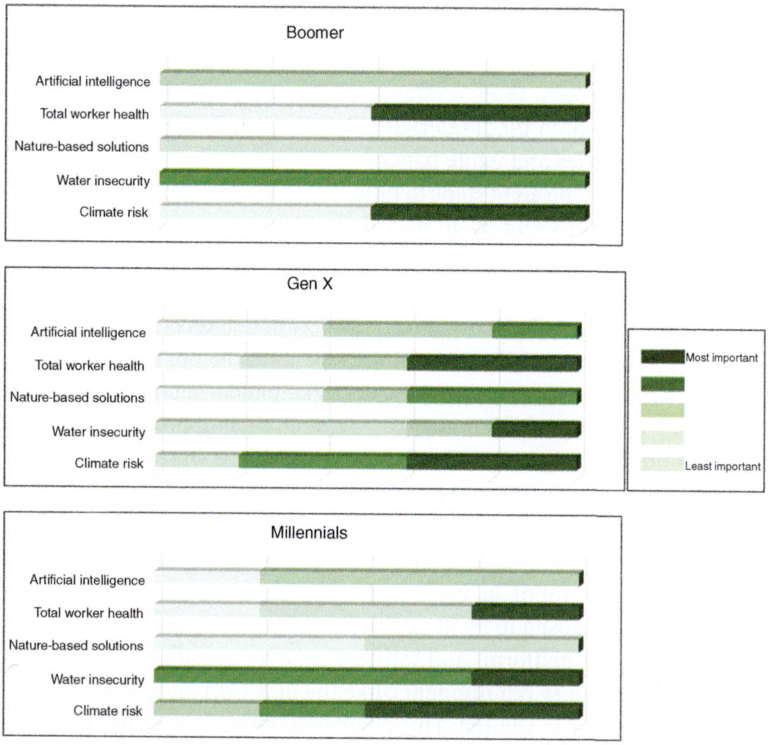

Figure 12.4 The generational materiality ranking within the mutual materiality experiment. Source: Woodard & Curran / with permission of Woodard & Curran.

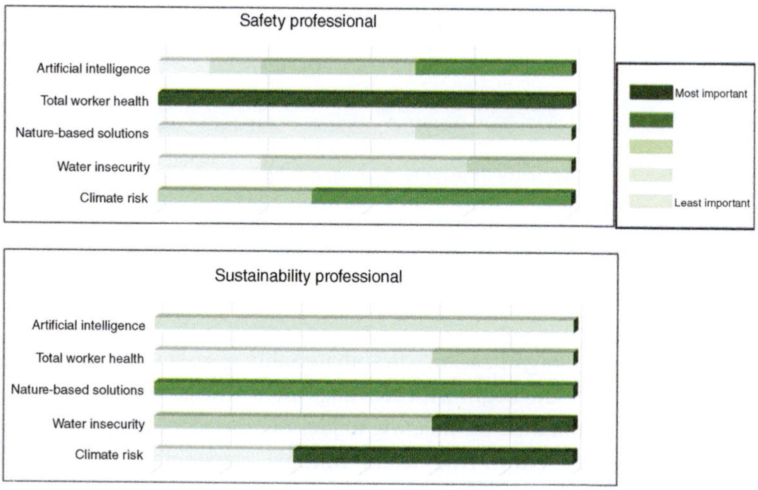

Figure 12.5 The profession materiality ranking within the mutual materiality experiment. Source: Woodard & Curran / with permission of Woodard & Curran.

The Materiality Process

The materiality process is also called the materiality assessment. An internet or ChatGPT search for "Materiality Process steps" will return multiple steps, links to articles and white papers.

The authors provide eight steps to fully stretch the process into its component parts to bring visibility to each individual part to show how each applies and takes shape. It is important to note that the order in which the organization completes the steps may reflect the organization's resourcing, maturity, or structure. Some organizations may have a set of key issues from past implementation years or determine key issues before engaging stakeholders whereas others may engage select stakeholders before finalizing the key issues – each organization will be unique in their approach since there is no set right or wrong order. What is most important is you are on your way!

There are numerous software platforms available that support the materiality process. These systems can complete and document much of the assessment for organizations – either through partnering for input at each step or with little input at each step, depending on how involved the organization has chosen to be in establishing the platform implementation engagement. Some platforms and software are even specifically designed to meet newer regulatory requirements, like the new Corporate Sustainability Reporting Directive in the EU. This chapter is written to "show the steps" independent of software systems to help junior practitioners better understand the process and explore what technology has to offer and what best enables success at your organization. It is certainly ever-changing!

See Figure 12.6 "Materiality assessment process steps (Marquette 2024)."

Figure 12.6 The materiality assessment process steps.

Step 1 – Goal Setting (Otherwise Known as the Purpose and Scope)

Whether this is the organization's first materiality assessment or its tenth, each assessment should begin with goal alignment. The goal alignment process may be a formal kick-off meeting to identify the scope, resources, and boundaries for the effort (the team, funding, geographic region, and business applicability) and timeline, or the process may be no more than to sign off on an annual or bi-annual purchase order. However, the authors would caution practitioners to strive to follow the formal process at a minimum to document decision-making for future professionals coming into the team.

In addition to being able to understand historical efforts to reduce time and resources by not duplicating prior efforts, having a historical record strengthens foundational programs by recognizing and building upon championed and prior successes while avoiding challenges and hurdles previously identified. Sustainability implementation within an organization is at its core about changing culture. While there will be champions within each level and function of the organization, each function and management level may also be at a different stage of maturity – each with its own successes, champions, nay-sayers, and hurdles. Therefore, any opportunity to get a step ahead should be viewed as indispensable.

Organizations may choose to conduct a material assessment:

- Create a more dynamic corporate strategy
- Refine resource allocations
- Prioritize actions, goals, and targets
- Engage with stakeholders toward alignment
- Reduce risk and leverage opportunities
- Determine disclosure topics
- Meet framework requirements

Materiality can inform the corporate strategy by identifying key issues, that is sustainability-related topics, of importance to stakeholders or perhaps reprioritizing issues previously on the company radar. As global trends impact local decision-making, prioritization may change year-over-year.

The materiality process can also be used to identify the relevant sustainability focus areas that may lead to or are linked to projects and initiatives aligned with the evolving corporate sustainability strategy, thereby refining resource allocations. As the stakeholders are identified and engaged, projects and initiatives are captured, and goals and targets may be set, sunset, or reset based on the changing priorities of the stakeholders. As each of these elements is incorporated into the corporate or organizational sustainability strategy, the potential to reduce operational risk and leverage opportunities toward competitive advantage increases.

The materiality assessment provides the initial outline for external disclosure topics, which can be crafted to represent single (financial), double (financial and impact), or triple (financial, impact, and context) materiality.

Step 2 – Stakeholder Identification

In the materiality assessment process, like benchmarking, the organization should identify their key stakeholders (see Chapter 10 for more discussion on stakeholders and stakeholder mapping). Stakeholders may include:

- Investors
- The board of directors
- The executive team (a.k.a. the C-Suite)
- Workers
- Community
- Suppliers
- Customers
- Consumers
- Competitors
- Peers
- Academics
- Regulators
- Nongovernmental agencies
- ESG rating agencies

For the materiality process, the organization may choose to engage some or all the stakeholders depending on levels of previous engagement or alignment with the stakeholder group, resources available to invest in the engagement strategy, and overall goals of the materiality assessment. To reach a broader group of stakeholders, the organization may employ different engagement methods for each stakeholder group. The team may interview the leadership team but electronically survey customers and suppliers. The materiality team, identified in Step 1, will map each stakeholder group to the qualitative or quantitative methodology planned to assess their material views. The engagement strategy may change from year to year based on previous information learned.

The Step 2 output will be the creation of the initial list of anticipated stakeholders the organization will be engaging as part of the overall materiality assessment. As the materiality process proceeds, this list should adjust depending on the issues that are identified, the conversations and interviews that are had, and the stakeholder concerns that are raised.

Step 3 – Benchmarking

Just as outlined in Chapter 11, within the materiality process, benchmarking similar organizations can provide a window into sustainability issues that are important to similar stakeholders and the related value chain, including competitors, peers, suppliers, and customers.

The benchmarking activities in this case may be a desktop literature search, that is, a website review, to collate materiality matrices from select companies, or it may include interviews with key stakeholders. The level of rigor will be based on the organization and the overall goal of the benchmark assessment. Refer to Chapter 11 for further guidance on conducting a benchmarking assessment. In this case, the peer stakeholder information is collated from websites to inform the organization as to what peers have deemed to be material issues. However, in this case, what may not be visible, without a transparent sustainability report by each peer company, is which stakeholders those peer companies considered when determining their materiality. Thus, peer benchmarking information is one series of data sources amongst a variety of stakeholder sources.

Step 4 – Key Issue Determination

Determining the key issues to include in the materiality engagement is gathered from various sources, including benchmarking efforts, internal and external stakeholder engagements, and various voice of customer exercises. Many industry associations also highlight key issues impacting the sector. There are white papers, articles, and periodicals published that provide insights into relevant issues and topics for organizations to consider. There are options for engagement at every level of resource and timing.

As previously reviewed, the International Financial Reporting Standards (IFRS) organization oversees sustainability accounting standards through the International Sustainability Standards Board (ISSB). The ISSB has absorbed the 77 sector-specific material disclosure standards originally drafted by the Sustainability Accounting Standards Board (SASB). The SASB Materiality Finder lists general issues across five broad categories: environment, social capital, human capital, business model and innovation, and leadership and governance (IFRS Foundation 2024). These five broad categories apply to nearly all organizations in some manner of risk and/or opportunity which are detailed in their sector-specific standards, which can provide an initial key issues list for organizations to consider. In fact, if the organization is preparing its materiality assessment as part of a disclosure effort for its investors, it may specifically choose to use the ISSB's two overarching voluntary standards that apply to all organizations and then choose SASB's appropriate sector-specific standards as an initial guide. The ISSB disclosures identify key issues that have the potential to financially impact the investment, more commonly known as *single* materiality as we discussed earlier in this chapter.

The GRI Standards include universal disclosures that apply to all organizations, sector-specific guidelines, and then topic-specific standards, like health and safety. The GRI Standards are drafted to identify impacts (key issues) not only impacting the organization but also how the organization conducting business impacts the environment and the community stakeholders, more commonly known as *double* materiality. And, while the community may seem a narrowly defined stakeholder, the term stakeholder has broadened to encompass nearly all potentially impacted stakeholders (BSR 2024). This broader view of stakeholder is reflected in their sector-specific guidelines (GRI 2024). When complete, the GRI will have completed 40 sector-specific guidelines.

As previously noted, few organizations are disclosing based upon context-based materiality; however, three leading examples stand out – Unilever, Kering, and IKEA (IKEA 2023; Kering 2019; Unilever 2024). In these examples, you may not be able to see the detailed work behind their materiality or key issue determination; however, you are able to read how the companies have tied their sustainability journey to the larger planetary story.

Ultimately, regardless of whether the organization chooses single, double, or triple materiality, coming out of Step 4, the list of issues should be prepared alongside the stakeholders for engagement in Step 3.

(At the time of printing, MSCI, a leading ESG Ratings agency, published its new website showcasing its industry materiality profiles. The profiles incorporate more than a decade of information from their global portfolio of companies and updates based on new information from issuer scoring; thus, as they remain publicly available, they are a dynamic representation of material issues across MSCI's evaluated organizations. The materiality map can be found at https://www.msci.com/our-solutions/esg-investing/esg-industry-materiality-map. While providing valuable information on issuers within the MSCI portfolio, users of their materiality map will want to review their process to understand how MSCI has defined theirmateriality process.)

Before we jump into Engagement, and for ease of explanation, we will assume, we are self-defining our key issues list in Step 4 from references defined via SASB GRI, or benchmarking, etc. For future documentation purposes, whether to include in the sustainability disclosures (report) or your internal materiality files, create a table delineating each stakeholder group that has been selected for engagement in Step 5, the engagement

Example Stakeholders			
Shareholder	Workers	Suppliers	Trade associations
Type of Engagement			
Annual meeting	Internal surveys	ESG surveys	Trend reports
Earnings meeting	Communications	ESG Audits	Trade KPIs
Regular meetings	Worker meetings	Supplier meetings	Trade surveys
Issues to Discuss			
Governance	Climate change	Packaging	Animal welgare
Climate change	Health and safety	Food waste	Regenerative Ag
Human rights	Food scarcity	Human rights	Nutrition and wellness
Innovation	Transparency	Innovation	Food safety

Figure 12.7 Stakeholders and materiality chart.

method, and topics to be discussed. Noting that not all material issues will hold the same relevancy to all stakeholders.

See Figure 12.7 "Stakeholders and materiality engagement chart (authors' rendition)."

Step 5 – Engagement

Depending on the organization, this step may lead to a list of material issues, or it may begin the prioritization process.

The engagement step sets out and defines the interaction between the organization and the identified stakeholders. It may be as informal as listing the stakeholders to include in the survey, or ideally, it is a fully flushed communication plan paralleling the table written in Step 4. The organization may engage with both internal and external stakeholders as part of the materiality process. The types of engagement may be informal, with or without the use of a software platform, and include:

- The annual meeting of the stockholders
- Quarterly earning meetings
- Customer and supply chain surveys
- External stakeholder surveys (NGOs and community)
- Employee engagement surveys

- Internal communications platforms
- In-person interviews
- Industry conferences

Whether the organization chooses to conduct interviews with leadership or survey the stakeholders using questionnaires, understanding the pros and cons of using both techniques is helpful. The interview allows for open-ended questions, which allow for free-flowing conversation. This does, however, remove anonymity between the interviewer and interviewee. Questionnaires have been referenced as being free from bias, but this may not always be the case based on how terms are defined and presented within the survey (Akbayrak 2000). Ideally, surveys are more cost and time-effective to execute for a much broader stakeholder group. On the other hand, the interview would generally have a higher response rate. As organizations grapple with the challenges of executing the materiality assessment, they may revisit their first step and refine their scope as they learn more about their stakeholder groups and the materiality process. As previously noted, having a third-party partner, whether it is the software vendor, platform support, or a consultant, may be helpful in navigating the process, but it is certainly not a requirement.

There are surveys asking the stakeholder to rank the issues from 1 to 10 in order of importance, others confirming whether an issue will continue to be an investment issue this year; the authors have also seen questionnaires asking the respondent to choose one topic over another. There are many

Figure 12.8 Sustainability questionnaire example.

types of questionnaires and surveys, just as there are as many interview questions to ask. Organizations will select the format that is most appropriate for the organization at the time of their efforts and likely improve upon the process as their sustainability program matures.

See Figure 12.8 "Materiality questionnaire example (authors' rendition)."

To accompany the engagement strategies, refer to Communication Plan Sam.

> The interview may be conducted as three open-ended questions:
> - What is most critical to the sustainability success of our organization?
> - What do you see as our greatest sustainability risk?
> - What is our greatest sustainability opportunity?

Communication Plan Sam is a fun way of thinking through how to create the right message and medium to develop not only the internal marketing communications plan but also how Sam can be adapted for the purposes of thinking through how to engage the stakeholders through the right method to get the best input on materiality. What level of engagement is desired for each stakeholder group? What is the key to developing a longer-term relationship? Consider a communication medium and cadence that allows gathering of material issues information but also to provide feedback to stakeholders. See Chapter 10 for a more in-depth discussion on engaging with stakeholders.

1) Objective
2) Context
3) Target audience profile
4) Messengers
5) Mode of communication
6) Media-specific strategy
7) Message for each unique audience
8) Call-to-action
9) Cadence
10) Monitoring KPI

See Figure 12.9 "Communications plan sam (Marquette 2024)."

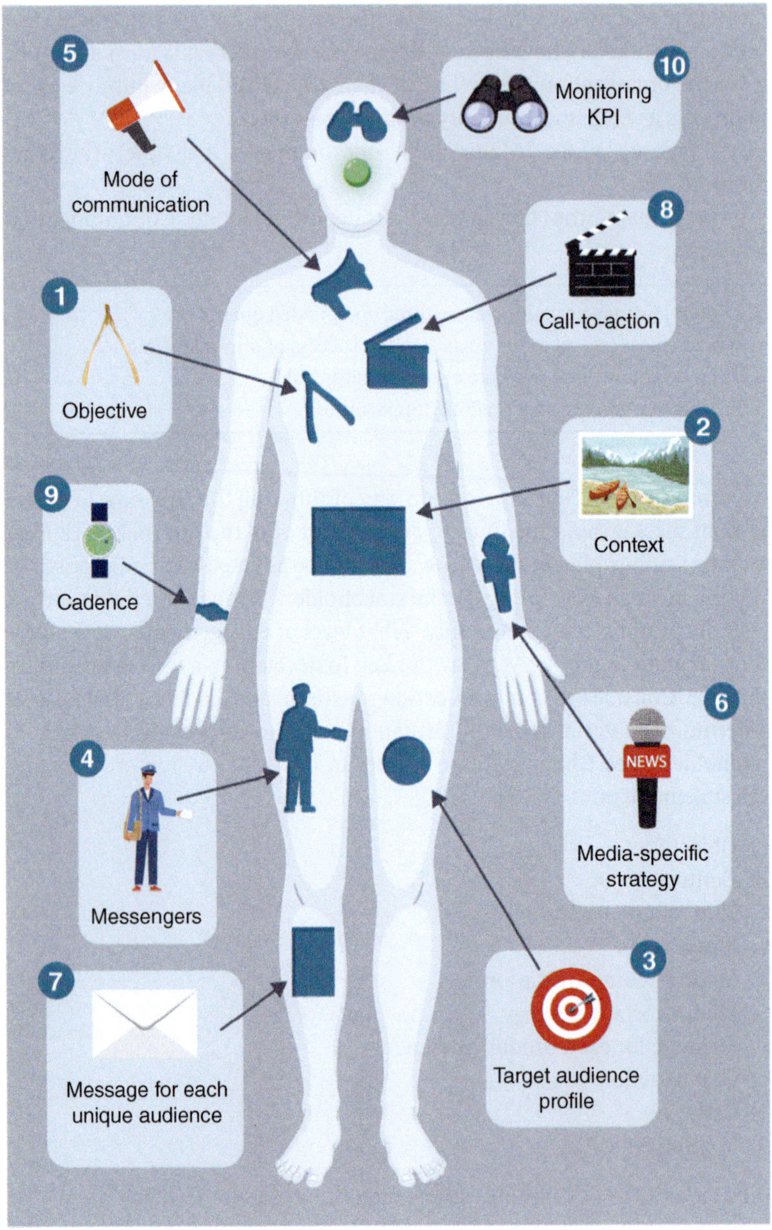

Figure 12.9 Communications Plan Sam.

Step 6 – Analysis

Both the qualitative data and quantitative results provide valuable insights comparing data from each stakeholder group and across stakeholder groups, internally and externally, and at different levels of the organization, all of which guide the sustainability disclosures strategy and materiality results. Leadership interviews, voice of customer interviews, survey results, and qualitative survey word count analysis – all can be useful means to understand the material issues identified in earlier steps, and likely some that were not. Equally as valuable is the comparison of stakeholder group responses from year to year to understand what is driving their attention as the market changes.

Examples of qualitative data analysis would be leadership quotes and direct responses that could be formulated to tell the strategic story and combined with quantitative prioritization of issues set the direction of the sustainability initiatives. Qualitative data may also be represented by word count, that is, literally by a count of each material issue identified by key stakeholder. You may have seen a similar representation as qualitative data word counts, which are often shared as "word clouds", with the higher word counts represented by larger font sizes and different variations on different terms, perhaps in different colors or font styles.

The output of the materiality analysis step allows both the qualitative and quantitative data to be prioritized in groupings or rank order of how both the organization and the stakeholder perceive the material issues.

Step 7 – Prioritization

Many organizations choose to represent the output of their materiality graphically with high, higher, highest ranking, or no order of ranking at all, while others will share material issues in alphabetical order without prioritizing the issues. Admittedly, it is rare to see materiality matrices with high-medium-low priorities as organizations would prefer not to explain how a sustainability issue is a low priority, as the authors have experienced multiple times. A simple internet search of "materiality matrix" on the image tab will yield a plethora of different styles of materiality representations. While many organizations do use the simple four-block, others may use 9-blocks, concentric waves, alphabetical lists, or many other creative, interactive disclosures that provide information on key issues by stakeholder group (often

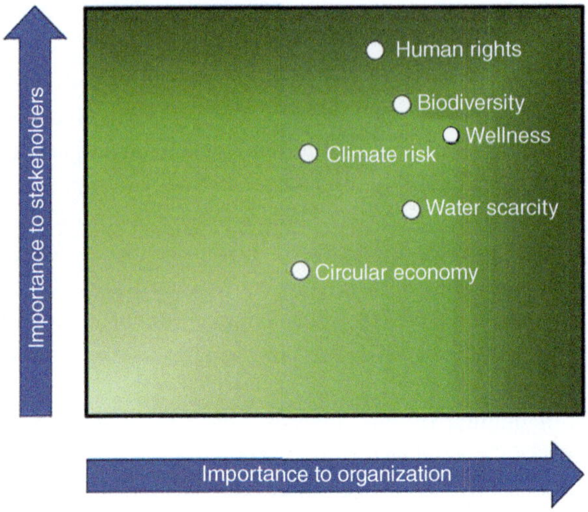

Figure 12.10 Simple materiality matrix.

represented as the organization and the stakeholders). The simplest form is below.

The closer an issue is to the top right, the more important, material, or strategic the issue is for the organization's sustainability strategy.

See Figure 12.10 "Simple materiality matrix (authors' rendition)."

Step 8 – Activation

Activation is the hard work that comes after the materiality matrix has been completed as the organization works to fully integrate the prioritized list of material issues across the various functions' programs, practices, and procedures. If the materiality assessment was completed as part of an initial strategy exercise, integrating the prioritized material issues into the sustainability strategy will likely align with the preliminary initiatives, goals, and targets.

As the organization strives to incorporate sustainability into existing management systems, remember sustainability is not a standalone siloed activity. It is not another new or separate task. It is just an additional perspective, an enhancement to the great work already underway. When endeavoring to introduce sustainability into a department, build on processes that are working – processes that have been accepted and the teams have fully adopted. Work with the team to introduce a sustainability lens (or layer) to their process.

For example, in process or project engineering, stage gates are used to filter projects for Go-No-Go decisions based on set criteria to provide predesigned opportunities to move forward or stop the project from unnecessarily wasting resources – the earlier, the better. Certain questions must be answered to move to the next stage. Health, safety, and quality questions are often included, for process safety-covered facilities, at each stage gate. Sustainability questions can be included in the same manner.

See Figure 12.11 "Engineering process incorporation of sustainability principles."

By building on existing successes, increases early adoption of sustainable practices and reduces the time to realization of opportunities and sustainability-related risk reduction.

Who Is Generally Involved?

The materiality effort generally involves practitioners from across the organization and may be unique for each company depending on where champions exist and the current level of sustainability maturity. However, ideally, representation from each department will ensure a robust strategy is developed that reflects each function and provides opportunities for all workers' perspectives to be incorporated into the engagement. This may mean the team is led through EHS and HR or perhaps through the CFO's office. Most organizations will strive to create a sustainability committee, a cross-functional team, to coordinate their efforts and work across the organization to introduce opportunities to collaborate with and seek input from each function.

Where Is It Interconnected?

Materiality sets the tone for sustainability efforts, whether conducted in-house, with partners, or through third-party collaborators. Sustainability is one of the rare topics that intertwines throughout the organization and draws teams into a picture on a single canvas. As the organization continues its journey, its materiality effort will evolve with it. Stakeholder engagement may evolve into its own outreach effort becoming a unique aspect of its strategy.

Integration into Other Strategic Elements (Chapters)

Over time, two key elements of the sustainability strategy pair to set the tone for the overall strategy – the stakeholders and materiality. The two

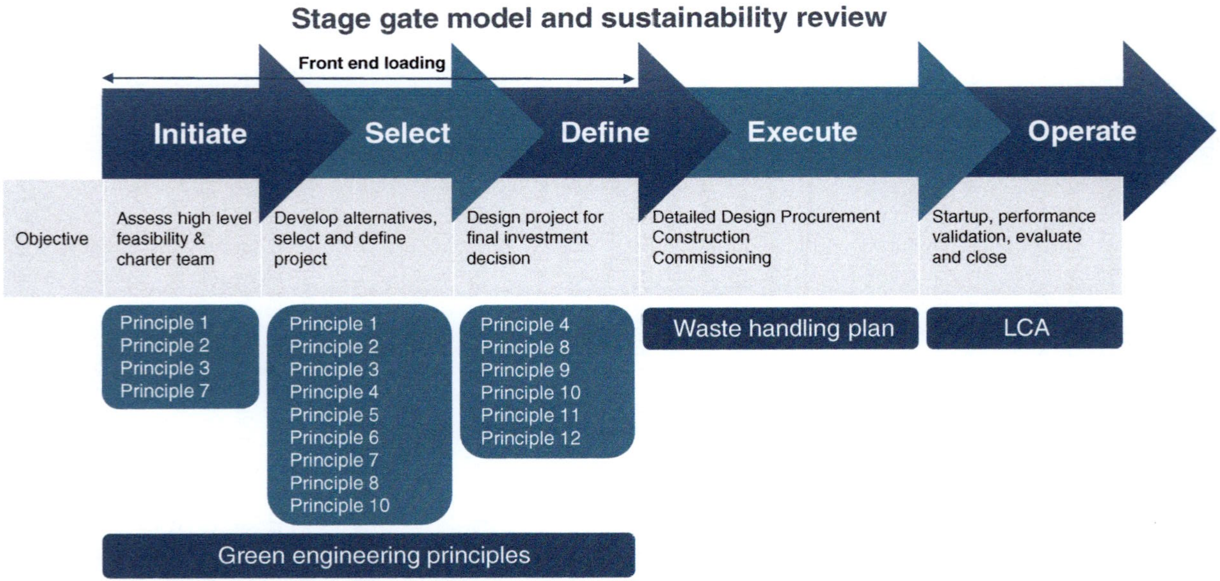

Figure 12.11 Engineering process incorporation of sustainability principles.

could not be more intertwined. The authors have worked with partners that have begun their journey with stakeholder analysis and others with a foundation of materiality, as the saying goes . . . it all comes out in the wash.

Operationalizing sustainability is not unlike the transformation environmental health and safety leaders have led and continue to lead. Having a passion for and dedication to environmental protection and the health, safety, and well-being of workers has been at the forefront of the profession for over a century. The materiality assessment, coupled with stakeholder analysis and benchmarking, within the Integrated Sustainability Business Model provides the environmental health safety and sustainability professionals and practitioners with more tools in their toolbox to advance their mission and bring competitive advantage to organizations.

References

3BLMedia (2024). GRI and IFRS collaborate to enable seamless sustainability reporting 28 May 2024. https://www.3blmedia.com/news/gri-and-ifrs-collaborate-enable-seamless-sustainability-reporting (accessed 25 February 2024).

Akbayrak, B. (2000). A comparison of two data collecting methods: interviews and questionnaires. *Hacettepe UniversityEgitim Fakultesi Dergisi* 18: 1–10. http://efdergi.hacettepe.edu.tr/yonetim/icerik/makaleler/1051-published.pdf

BSR (2024). FAQ stakeholder engagement. https://www.bsr.org/en/prs/stakeholder-engagement (accessed 25 February 2024).

GRI (2021). The double-materiality concept application and issues. https://www.globalreporting.org/media/jrbntbyv/griwhitepaper-publications.pdf (accessed 25 February 2024).

GRI (2022). IFRS foundation and GRI to align capital market and multistakeholder stakeholder standers 24 March 2022. https://www.globalreporting.org/news/news-center/ifrs-foundation-and-gri-to-align-capital-market-and-multi-stakeholder-standards/ (accessed 25 February 2024).

GRI (2024). GRI sector program. https://www.globalreporting.org/standards/sector-program/ (accessed 25 February 2024).

IFRS Foundation (2024). SASB standards materiality finder. https://sasb.ifrs.org/standards/materiality-finder/ (accessed February 2024).

IKEA (2023). People & planet positive. https://www.ikea.com/us/en/files/pdf/19/ac/19acbc74/ikea_sustainability_strategy_2023.pdf (accessed 25 February 2024).

Katz, D. and McIntosh, L. (2021) Corporate governance update: "materiality" in America and Abroad. Harvard Law School Forum on Corporate Governance. https://corpgov.law.harvard.edu/2021/05/01/corporate-governance-update-materiality-in-america-and-abroad/ (accessed 25 February 2024).

Kering (2019). Linking planetary boundaries to business. https://www.kering.com/en/news/linking-planetary-boundaries-to-business/ (accessed 25 February 2024).

Marquette, T. (2024). Understanding materiality and risk. PowerPoint Presentation, National Safety Council Future of EHS Conference Louisville KY, 16 February 2024.

McElroy, M.W. (2019). United Nations research institute for social development making materiality determinations a context-based approach. https://cdn.unrisd.org/assets/library/papers/pdf-files/wp2019-6-mcelroy.pdf (accessed 25 February 2024).

Reuters, T. (2022). Financial materiality: understanding the financial performance of ESG strategies 7 September 2022. https://www.thomsonreuters.com/en-us/posts/investigation-fraud-and-risk/financial-materiality-esg/ (accessed 25 February 2024).

Sustainability Accountability Standards Board (2016). Letter to Mr Brent J Fields, Secretary, United States Securities and Exchange Commission dated 1 July, 2016. https://www.sec.gov/comments/s7-06-16/s70616-25.pdf (accessed 1 July 2024).

Unilever (2024). Leading in the next era of corporate sustainability. https://www.unilever.com/sustainability/ (accessed 30 June 2024).

US Supreme Court (1976). TSC Industries, Inc. v. Northway, Inc. 426 US 438 https://supreme.justia.com/cases/federal/us/426/438/ (accessed 20 June 2024).

13

Value: Creation, Assessment, Valuation, and Impact Pathways

An Introduction to Value Assessment

In the context of sustainability and the "Capitals", value assessment is a *Sustainability-Related Assessment and Implementation tool (SAIT),* originally introduced in Chapter 9 *The Sustainability Strategy and Decision Making Roadmap*. The overarching objective of value assessment is to inform decision-making that ultimately improves company performance while creating value for all stakeholders (Capitals Coalition 2016, 2019, 2022a, 2022b; Impact Management Platform 2024). It is important to note that value for a company and its stakeholders can be created or eroded depending upon the company' activities (Capitals Coalition 2022a, 2022b).

What is value in the context of sustainability? It is a measured result of identifying and assessing a company's impacts and dependencies through a valuation process such as an impact pathway. Impact pathways are described in more detail later in this chapter. Measurement of impact and resulting value created or eroded can be described in quantitative, qualitative, or monetary terms (Capitals Coalition 2022a, 2022b). Once a company determines the true value of a its impacts and dependencies on people, business, nature, and society, this value can influence company decisions on business strategy, including prioritization, resource allocation, accountability structure, impact risk mitigation, and communication with stakeholders.

Value-based decision-making can be internally focused on business strategy, mergers and acquisitions, action planning, communication, transparency, accountability, human resource allocation, capital expenditures, and resource allocation, day-to-day operations, etc. While external value-based decision-making can emanate from impacting, depending upon or influencing stakeholders such as customers, suppliers, government policy

Connecting the Dots: Environment, Health, Safety, and Sustainability,
First Edition. Kathy A. Seabrook and Tanis J. Marquette.

makers, investors, rating agencies, insurers, regulators, employees, and other workers. In both instances, decisions are influenced by whether value is created or eroded by company activities. Identifying, measuring, managing, and communicating a company's current and future impact and dependency risk is at the core of value assessment and resulting decision-making.

Throughout this book, we have used an expanded capitals approach to understanding company value. To this end, in addition to measuring and accounting for traditional financial capital in a company's profit and loss reporting, this capitals approach also incorporates the value created by a company's activities on human, social, natural, and produced capitals. This "Integrated Profit and Loss" (IP&L) report and methodology can play a key role in identifying and understanding the interconnectivity between social, environmental, and human capital impacts and dependencies, how they relate to one another and financial performance. (Natura & Co 2022) Value creation in this context may include impacts and dependencies on employees (human capital), families, communities, and business relationships (social capital), preserving the natural environment (natural capital as well as a company's assets, products and services produced (produced capital) as a result of company activities. Chapter 6 *The Capitals* provides a deeper dive into this capitals approach to value creation. A company's operations and activities (i.e. monetizing risk reduction improvements), its advantage over competitors (i.e. increased market share and societal value created by its products and services) and business relationships with customers, suppliers, and its workforce create value and this value can be monetized in many instances. Let us look at social capital in the context of social and societal value.

Social and Societal Value

According to Angus Bauer, Head of European Equity Research at Schroders (asset managers), "the social value of connectivity was one of the working groups that I led where we try and actually put a dollar number to the societal value of digital connectivity, the value that goes unpriced." He goes on to say about this digital connectivity through a mobile phone example, "one may pay US$150 per month for a mobile (cell) phone service with unlimited data, domestic phone calls, etc. Does that mean that the total social value associated with the value chain that produces this service is capped at US$150 a month or is there a lot more utility that is unpriced and is effectively surplus value created by the mobile (cell) phone service provider and the industry?" (Bauer and Seabrook 2023) This positive externality of free digital connectivity for customers and society can be valued, monetized, and expressed as a percentage of sales to normalize for the size of the company. This is social capital for the company (Bauer and Seabrook 2023).

Bauer puts this in perspective. "At the end of a given year, one can look at its business and surmise which would be true: the world would either give it a bill to pay for the social cost of running the business that is not borne by the balance sheet or profit and loss reports, or provide a check for the social benefit that does not accrue to the business" (Bauer and Seabrook 2023). The difference is the social cost or benefit from the social capital accrued or eroded. In the case of the mobile (cell) phone company, it would receive a check from the social capital realized by providing free digital connectivity. This accounting ultimately provides a truer account of productivity or profitability of a business (Bauer and Seabrook 2023). See Table 13.1 for an in-depth case study on digital connectivity and the social value it can create.

Because impacts can be positive or negative, they, along with dependencies, can be mitigated if a negative impact or dependency is the outcome of assessing the value of an activity. These mitigations can be measured to demonstrate how they create value for a company, its stakeholders, and the natural environment, including the people that make up its workforce. The value of these mitigations as well as the value of the resulting positive impacts can then be taken into account in decision making by the company and its stakeholders on whether to implement those mitigations.

Communicating Value: Natura's IP&L Report

A more comprehensive accounting of the company's true value is captured by integrating valuation of all the capitals in reporting (Capitals Coalition 2016, 2019, 2022a, 2022b; Natura & Co 2022; National Safety Council 2023). Specifically, accounting for the impact of corporate activities on human, social, and natural capitals, measuring and, where possible, monetizing the resulting value created provides opportunities to communicate this value through traditional profit and loss accounting.

Natura published their first IP&L report in 2021. Natura is part of Natura & Co, which is the fourth-largest beauty company in the world (Natura & Co 2022). In addition to Natura, Natura & Co includes Avon, The Body Shop and Aesop brands. According to *Sustainable Business Now*, Natura believes the future of finance will reflect environmental and social impacts and opportunities in financial reporting. According to Denise Hills, Natura & Co Latin America's sustainability director, their "IP&L provides a way of

Table 13.1 A case for social value: a mobile (cell) phone company.

A case for social value creation: a mobile (Cell) phone company
Let us use an example of a mobile (cell) phone company to better understand social value.
One can have a variety of apps on their mobile phone. A navigation app can ensure that one never get lost. One globally recognized messaging app currently on the market can act as a "social networking tool" by having 95% of one's friends, family and colleagues on the messaging app, allowing free connectivity with them via text or a telephone call across the world. Use of this app means one can organize to meet up or communicate with one's social network, separately or together, at a moment's notice and for free. This is unpriced social capital for the mobile phone company.
On the personal side, hotels are now using this same globally recognized messaging app to provide instant concierge service. This means quick access to restaurant recommendations and instant requests and delivery by the hotel's room service.
One can also do business through this globally recognized messaging app, talking to data or information providers or if a retail provider, sell things through the messaging app. Theoretically, the app can provide an entirely free channel through which to reach customers. This creates value for the company, its customers and its network of users.
These examples are a positive consequence of the free connectivity through the apps. This is social value to both the phone owner and the service provider. However, this social value is not typically accounted for on the mobile phone service provider's balance sheet.
To further capture this idea of social value, think about how much you would want the mobile phone service provider to pay you to give up these apps or to rescind all access to your phone. Would you expect to give it up for less than the US$150 a month you pay for the mobile (cell) phone service? The monetary difference in the monthly cost for service and the value you put on the benefits for free use of the apps is the social value created for you, a customer (an externality to the company) and ultimately the mobile (cell) phone service provider (an internality).
Sustainability seeks to value externalities in addition to internalities. Whether it is the social cost of carbon emissions (a negative externality to the company) or alternatively, the social value created by free digital connectivity via a mobile phone service provider, a positive externality., both can be accounted for in an IP&L and influence decision making.

Source: Bauer and Seabrook (2023).

Table 13.2 Societal benefit (value) of Natura's Amazon initiative.

For the extractive communities (Natura's value chain)	Brazilian real$(millions)/approximate US$(millions)
Natural Capital (i.e. regenerative forestry, extraction and biodiversity)	R$21.6 (US$3.9 m)
Social Capital (i.e. extractive communities/ingredient suppliers)	R$287.5 m (US$51.4 m)
Human Capital (i.e. personal income generated)	R$55.7 m (US$10)
Total	R$317 m (USD$56.7 m)

Source: Adapted from Natura and Co (2022).

measuring impact and showing a fuller view of economic value beyond finances. It was a strategic decision. Every company discusses costs but not many really understand the associated value" (Sustainable Business Now 2024).

in 2000, Natura launched its Ekos brand, using ingredients from Brazilian biodiversity in the Amazon. This started the Natura's relationship with Amazon extractive communities that use a regenerative model, known as agroforestry, to produce key ingredients for Natura's products. This regenerative model incorporates forest preservation and restoration with a focus on the health and well-being of soil and the environment. This in turn protects the biodiverse ecosystems that rely on the forests, including wildlife and plants. In their 2021 IP&L, Natura captured the societal benefit (value) of this initiative for the Amazon extractive communities in their value chain. See Table 13.2 for a breakdown of the monetized societal benefit (value realized) by this Amazon extractive communities initiative (Natura & Co 2022).

Value Creation: "The Sweet Spot"

Dave Stangis is a seasoned chief sustainability officer and Katherine Valvoda Smith is the executive director for the Boston College Center for Corporate Citizenship in the Carroll School of Management. Figure 13.1 is reprinted from their book, *The Executives Guide to twenty-first Century Corporate Citizenship.*

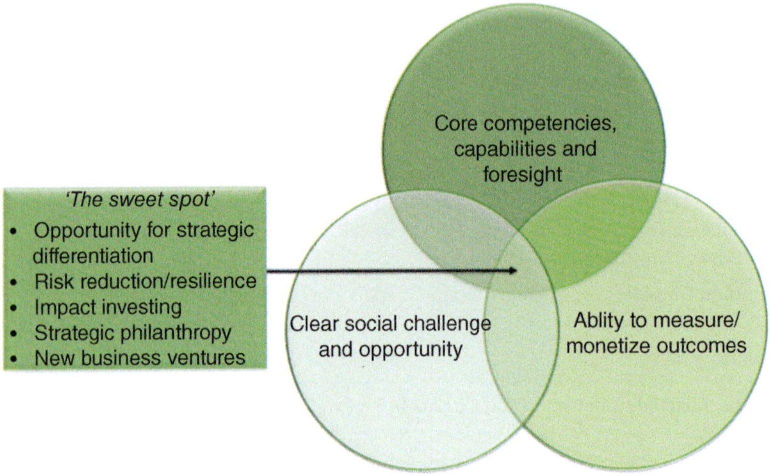

Figure 13.1 Value creation – "the sweet spot": philanthropy example.
Source: Stangis and Smith (2017)/with permission of Emerald Publishing Limited.

Figure 13.1 depicts the intersection between:

(1) a company's ability to identify, measure, and monetize its impacts and activities,
(2) the intentional, proactive company (purpose), capability, and impact activities, and
(3) a societal need or challenge.

This inflection point is the company's "sweet spot" according to the authors. It is where value is created for the company (Stangis and Smith 2017). When understood organizationally, this value creates opportunity and can influence executive leadership decision-making.

Valuation is the company's ability to measure and monetize its impacts and dependencies resulting from its activities and is an important factor for both the company and its stakeholders. Stangis' and Valvoda Smith's philanthropy example in Figure 13.1 depicts a mutuality of value created for communities, society, and the company. When a company uses its "core competencies, capability and foresight" to tackle identified community or societal challenges, such as nutrition, food scarcity, transportation, or public health, it is an opportunity for all impacted stakeholders. This "sweet spot" is where the company experiences the greatest tangible and nontangible value. Stangis and Valvoda Smith concur, when this value is monetized, it can be captured on the company's balance sheet to inform decision-making (Stangis and Smith 2017; Stangis and Seabrook 2023).

If we unpack this "sweet spot" from the philanthropy example, Stangis and Valvoda Smith suggest opportunities for value creation include (Stangis and Seabrook 2023; Stangis and Smith 2017):

(1) brand enhancement, differentiation, and competitive advantage,
(2) strengthened business relationships (i.e. community trust and human and social capital),
(3) risk reduction and resilience (influencing investor, insurer, employee, supplier, and customer decision-making),
(4) strategic philanthropy aligned with business capabilities and strategy, and
(5) potential for new business ventures (e.g. new products and services).

Expanding the applicability of the philanthropy model in Figure 13.1, human capital or natural capital challenges can be substituted for social challenges. For example, human capital challenges could be recruiting and retention, access to good jobs and career development and advancement, human rights, being paid a living wage, job satisfaction, worker health, safety, and well-being or productivity and efficiency losses and stifled innovation. Substituting human capital in the model produces a similar "sweet spot" (Stangis and Smith 2017; Stangis and Seabrook 2023):

(1) brand enhancement, differentiation, and competitive advantage,
(2) strengthened relationships with employees (human capital),
(3) health and safety risk reduction and better quality of life for employees and other workers,
(4) Improved employee retention and organizational resilience (influencing investor and customer decision-making),
(5) Business-aligned strategic talent management, and
(6) potential for new business ventures as a result of employee and other worker engagement and contribution to innovation.

The key to achieving this "sweet spot" is the company's ability to identify, understand, measure, manage, and monetize the resulting outcomes, impacts, and dependencies from its activities. It does this through its capacity to mitigate and manage social, human, or natural capital challenges, all aligned with the company's purpose, competencies, and capabilities.

This overview of value creation brings us to the process of valuation using a value assessment process, which we will discussnext.

Unpacking a Value Assessment

In the context of these capitals, regulators and other government officials, policy makers, investors, customers, suppliers, employees, value chain workers, nongovernmental (NGOs), and not-for-profit organizations are potential stakeholders and the business relationships (social capital) upon which the company depends.

Since a company's impact and dependency risks are a result of its activities, they can be positive or negative. In the case of the potential risks associated with a company's negative impacts or dependency on natural resources, people, customers, and value chain suppliers necessary to operate the business, they can be mitigated and managed. As mentioned earlier in this chapter, impact and dependency risk mitigation can be measured and, in many instances, monetized to demonstrate how implementing them can create value for a company and its stakeholders. These stakeholders include the people that make up the company's workforce. To provide more clarity on impact and dependency mitigation, Table 13.3 provides some specific examples. These mitigations are also known as impact drivers, which will

Table 13.3 Impact and dependency risk mitigation examples (aka impact drivers).

Impact and dependency risk mitigation (Impact drivers)
Examples: • initiatives to reduce greenhouse gas emissions: transitioning to renewable energy, reducing air travel, regeneration initiatives, and reducing energy use in operations; • initiatives to mitigate water dependency by reducing water use in operations; • initiating an apprentice program and other workforce development opportunities to prepare the future workforce • substituting scarce raw minerals in new product design for materials that are more readily available and less toxic to people and the environment; • regenerative mineral or raw material extraction in the Amazon – enhancing biodiversity preservation for plants and wildlife that depend upon the forests; • integrating corporate accountability systems such as audit functions inclusive of sustainability, ESG, and the capitals; • product safety initiatives; and • implementing ergonomic initiatives to reduce musculoskeletal injuries in the workplace. (This example will be discussed later in this chapter.)

Source: Global Solutions, Inc. and Seabrook (2024).

be discussed further in the Impact Pathways: Ergonomics Intervention Example section below.

The value created by mitigating impact and dependency risks may be influential in company and stakeholders decision-making relative to a company:

(1) engaging in forest land regeneration;
(2) implementing a water use strategy for product manufacturing; or
(3) initiating mandatory use of fall arrest systems when working at height across the company and its value chain.

Specifically, these impact risk mitigations can influence stakeholder decisions to invest in, work for, supply to, or buy from a company.

According to the International Foundation for Valuing Impact, "impact pathways provide a consistent method to measure impacts, allowing for comparability across time and between entities for a specific sustainability topic" (International Foundation for Valuing Impact 2023; Impact Management Platform 2024 ; Value Balancing Alliance 2023). What differentiates this value assessment process is it measures the value of a company's impacts and dependencies on people, business relationships, the natural environment, the economy, and society as a true measure a company's performance. Its #TrueNorth performance. In traditional financial accounting, value assessments strictly measure the impact of a risk to the company's financial capital. If communicated, understood, and embraced, this more holistic measure of a company's value and performance will influence decision-making on business strategy and action planning, including prioritization, resource allocation, accountability systems, impact and dependency risk mitigation, and communication with stakeholders. This approach of monetizing impacts using a monetary value factor can be accounted for on a company's balance sheet. Your finance and accounting teams are adept at valuation and an important ally to engage in human, social and natural capital valuation assessments. Their perspective and expertise will be invaluable.

Ultimately, value assessments and the decisions that ensue represent a measure of a company's true value by identifying the contribution of people, workers, stakeholders, and business relationships as well as the natural resources (the company's human, social, and natural capital) upon which the business depends (internalities, the risk to the company). The converse is also true. Value assessment is a true measure of the company's impact on stakeholders, natural resources, and human well-being. The Foundation for Valuing Impact and the Value Balancing Alliance also assert, "monetary valuation of impacts from the perspective of affected stakeholders refers to the estimation of the relative importance, worth, or usefulness of impacts to the people who experience the impact, expressed as a monetary value" (International Foundation for Valuing Impact and Value Balancing Alliance 2023). The OECD Framework and research also provides guidance for measuring well-being and progress that can be incorporated into value assessments (OECD 2024). It is important to note impacts can be experienced by people both directly and through changes to the economy or condition of the natural environment. These value assessments incorporate valuation on present and future impact and dependency risk and risk mitigation to the company's operations (International Foundation for Valuing Impact and Value Balancing Alliance 2023).

Impact Pathways

In the context of sustainability and the "Capitals," an impact pathway is a value assessment tool. It helps a company identify and measure its impact and dependency risk and risk mitigation activities on employees, workers (including value chain workers), families, communities, customers, suppliers, the business, the investment community, the economy, the natural environment, and societal well-being (Capitals Coalition 2022a, 2022b). The impact pathway draws upon an integrated approach to better account for the value of a worker's contribution to family, community, the business, and customers when they are injured or experience ill health as an example. This approach to valuation also focuses on the value created by community, customer, investor, government regulator, and supplier trust which supports brand integrity as well as the natural resources upon which a company depends.

Ultimately, impact pathways lay out a systematic approach that enables a shared organizational understanding of a clear, value-based methodology to valuation using integrated data on a company's impacts and dependencies. In many cases, realized value, when monetized, can enhance a company's financial and sustainability performance and reporting through new and existing internal and external communication mechanisms. Identifying and communicating the value created by positive impacts and mitigating negative impacts to all stakeholders is at the core of the impact pathway valuation process. This brings us full circle, to the overarching purpose of value assessment and impact pathways – to influence actionable decision-making.

Note to Reader: While the natural environment, including natural resources and biodiversity, are not typically considered a company stakeholder. There is a case to be made for the value of positive impacts on the natural environment. After all, it is the natural environment that supports human life and upon which a company depends now and into the future. This is where the dots connect between people, business, society, the economy, and the natural environment. It is a company's dependency on the natural environment, and this dependency can directly or indirectly impact the company, people, its stakeholders, society, the economy and the natural environment. The concept of dependency is discussed more fully in Chapters 4: *Theory of Change: Input, Outputs, Outcomes and Impacts* and Chapter 6 *The Capitals.*

Let us continue to unpack the impact pathway methodology using an Ergonomics mitigation example.

Impact Pathway: Ergonomics Intervention Example

In order to better understand how impact pathways are integrated into business decision-making, let us begin by exploring an ergonomics risk mitigation impact pathway shown in Figure 13.2. This example is adapted from the Capitals Coalition's Valuing Human Capital in Occupational Health and Safety white paper (Capitals Coalition 2022a, 2022b).

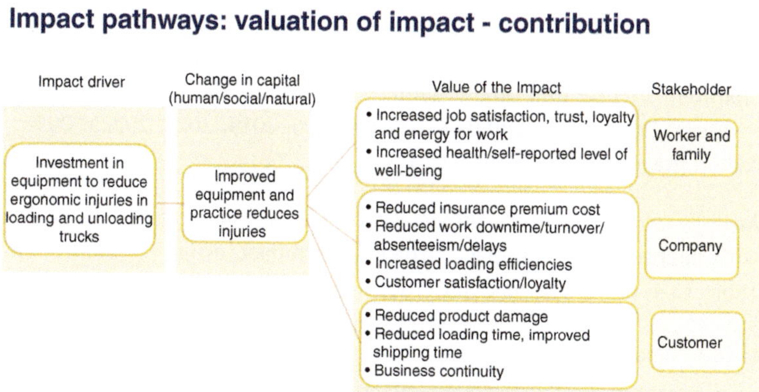

Figure 13.2 Impact pathways ergonomics example. Source: adapted from the Capitals Coalition work 2019, 2022a, 2022b; Global Solutions, Inc. and Seabrook 2024.

As we have discussed earlier in this chapter, an impact pathway depicts the relationship between a company's inputs, activities, and outputs, and their impact and potential dependency on people, business relationships, and the natural environment (Impact Management Platform 2024). An impact pathway demonstrates the business relevance of a company's impacts and dependencies. It provides a collaborative tool and valuation process, creating a company-specific business case for investing in human, social, natural capital, and other sustainability and EHS-related risk mitigation initiatives. It is important to note, risk mitigation activities such as an ergonomics, hearing conservation or crane safety program are considered drivers of impact (Capitals Coalition 2016, 2019, 2022a, 2022b). See Table 13.4 for the definition of an impact driver.

When an impact pathway is conducted collaboratively with cross-functional departments and stakeholder input, it enables a shared organizational understanding of the value created or eroded by a company's impacts and dependencies. Cross-functional departmental engagement includes finance and accounting, including the CFO, human resources, legal, investor relations and corporate communications, risk management, operations, research and development, EHS, sustainability, and other relevant team input. Engaging external stakeholders such as customers, suppliers, investors, etc., can also assist in the valuation process. Each of these internal and external stakeholders will have a lens through which mitigation of business risk and capitalizing on opportunities associated with impacts and dependencies will inform impact pathway outcomes, valuation and actionable, value-based decision-making. This ties action directly to the effects or consequences of the impacts, dependencies, and opportunities.

The following steps will help navigate the impact pathway process. They are adapted from the Capitals Coalition's *Valuing Human Capital in Occupational Health and Safety* white paper and in their Social, Human, and Natural Protocols (Capitals Coalition 2016, 2019, 2021, 2022a, 2022b; Impact Management Platform 2024).

1) *Identify and Measure an Impact Driver* – The impact pathway begins with identifying an impact driver. An impact driver is defined in Table 13.4. It is a business activity, an input or output, and is a catalyst

Table 13.4 Impact driver definition.

Impact driver
A measurable business activity, input, or outcome that can cause a change in a capital: positively or negatively, intentionally, or unintentionally.

Source: Adapted from Capitals Coalition (2022a, 2022b).

for change. It is a measurable quantity, a cost, or number of something such as people, trainings, or units of carbon emissions per miles traveled. The ergonomic intervention in the impact pathway example in Figure 13.2 is an example of an impact driver. Tables 13.3 and 13.5 provide additional examples of impact drivers.

2) *Measure the Change in the Capital(s)* – The next step in an impact pathway is to identify and measure the *change* in a capital(s) as a result of implementing the impact driver. This actual change in human, social, natural, produced, or direct financial capital can be positive and beneficial or negative, eroding any benefit of the change. Stakeholders are impacted by the change in capital, such as a worker and their family; the company's customers, suppliers, and communities in which they operate. Also, investors, society, government regulators, rating agencies, etc.

An improvement realized as a result of an impact driver (activity input or output), could be:

a) improved public health outcomes (social capital),
b) reduced worker injuries and ill health (human capital),
c) greater availability and access to natural resources (natural capital), and
d) access to financial resources for a capital expenditure (i.e. an acquisition of a company, purchase of equipment, or funding new research and development) (financial capital).

Table 13.5 Monetized impact drivers.

1) Cost of increasing renewable energy use by 30% per site (natural capital) or the number of sites increasing renewable energy use by 30% (natural capital),
2) cost associated with reducing greenhouse gas emissions or reducing the company's overall carbon footprint – by 40% over a five year period (natural capital),
3) cost to install a ventilation system to reduce hazardous chemical exposure to workers in a research and development lab (human and social capital) and the number of workers in the lab affected by the reduced exposure to hazardous chemicals (human and social capital),
4) cost of substituting toxic for less toxic raw materials on a chip manufacturing line or number of workers exposed to toxic materials on the chip line,
5) cost of investing in ergonomic equipment to reduce worker musculoskeletal injuries or the number of workers affected by implementing new ergonomic equipment (human capital),
6) cost of implementing diversity, equity, and inclusion policies and training, fair supplier agreements, or implementing an integrated EHS management system (human, social, and natural capital), and
7) cost of losing a key supplier due to unfair supplier agreements.

Source: Global Solutions, Inc. and Seabrook (2024).

For example, worker safety and health can be impacted by various business practices, policies, procedures, and other risk mitigation interventions. The health and safety of workers is an asset of human capital and impacts a worker's ability to contribute to the company. When worker safety and health is improved by a risk mitigation intervention, a beneficial change in the ability of a worker to contribute to the company is realized. This beneficial change also impacts the worker's family and the community in which the worker lives. Improving worker safety and health through risk mitigation initiatives (impact drivers) positively impacts overall worker wellbeing, family income, and quality of life. This creates human and social capital for the company. Forgoing health and safety risk mitigation and the company's dependency upon their workforce being fit for work can potentially impact the company's ability to produce quality products and services, reputational risk, brand integrity, and business continuity. This applies to a company's employees as well as workers in their value chain.

3) *Valuation of Impacts* – Assessing the value of an impact that is created by a change in a capital(s) is the endgame of collaboratively conducting an impact pathway. The value created can be expressed qualitatively, quantitatively or in monetary terms. Value creation results from impacts on the business, its stakeholders, the economy and society. Beneficial impacts to workers can include access to career development and equitable opportunities, public health, a living wage. Beneficial impacts to a company include access to capital, skilled labor, and natural resources. Financial, reputational, security, health, safety, well-being, access to resources, efficiencies, innovation, etc. are potential opportunities and benefits to a company.

The ergonomic impact pathway example in Figure 13.2 depicts a risk mitigation that drives improvement (a change) in human and social capital by reducing the potential for musculoskeletal injuries while loading and unloading trucks. The investment in ergonomic equipment is a risk mitigation and an impact driver in this example. It can reduce the potential for musculoskeletal injuries and the resulting financial burden on families increases worker well-being, retention, trust, and job satisfaction, strengthening a company's human capital (Capitals Coalition 2022a, 2022b). Among other benefits of the ergonomic equipment investment, workers' and their families' quality of life is also improved. Additionally, value is also created for the company, its brand integrity and relationships with its workers, the community, customers, suppliers, and society. From a societial standpoint, lack of decent work can lead to poverty, unrest and ultimately issues with the rule of law and security. This could impact a business' ability to thrive, ultimately resulting in site closures, disruption around product manufacturing, service delivery, business continunity and resilience (Capitals Coalition 2022a, 2022b).

Using an impact pathway valuation process goes beyond the traditional financial measurement of lost income and medical costs to the injured person or the cost of insurance to the company. For example, in an recent interview with a chief sustainability officer (CSO), this author was told the CSO's human resources director reported the cost for hiring an entry-level employee into the company was US$6000. Knowing hiring costs by job classification and the corresponding retention rates for those same job classifications results in the ability to monetize this human capital management metric. This example demonstrates why collaboration with peers is essential in value assessment. In this case, human resources provided the retention data to do this computation. The finance department could assist with accounting for the financial value of employee retention (human capital management) in the company's profit and loss reporting.

To monetize the value to the injured worker, the World Health Organization uses disability-adjusted life years (DALYs) to value the life lost or disability-burdened life lived. According to WHO, "one DALY represents the loss of the equivalent of one year of full health. DALYs for a disease or health condition are the sum of the years of life lost to due to premature mortality (YLLs) and the years lived with a disability (YLDs) due to prevalent cases of the disease or health condition in a population" (World Health Organization 2020). This concept allows a company to calculate the value of a worker's contribution in light of lost income, quality of life, mental health, ability to support family and the community, etc. For the business, the value is associated with the loss of worker trust, mental health of workforce, reduced access to knowledge, skills and abilities need for product manufacture, service provision, innovation, R&D and problem-solving. It is also the loss of that "discretionary energy" from the workforce as described by Paul O'Neill in Chapter 8 *Beginning the Journey*.

For in-depth DALYs calculations, the National Safety Council's *The New Value of Safety in a Changing World* research report provides examples of workplace safety-related computations defining and calculating value using this DALY methodology (National Safety Council 2023). See Table 13.7 for the results of a value assessment Nike conducted on monetizing the reputational risk of implementing a global supplier safety and health initiative.

It is important to note that while value assessment, and specifically this impact pathways process, seeks monetary valuation, that valuation is intended to provide a standardized approach for comparability of sustainability-related data at scale. Monetary valuation is not meant to put a cost on the intrinsic value of human well-being (Capitals Coalition 2022a, 2022b; World Health Organization 2020).

Other valuation considerations for the ergonomics intervention example are the insurance cost of worker ill health (human capital), cost for unscheduled production downtime in the event of an incident (human, social, and financial capital), and cost of reputational risk and brand integrity.

In general, valuation should include insurance, cost of worker injuries and ill health (human capital), cost for unscheduled production downtime in the event of an EHS or other incident (human, social natural, financial, and produced capital), cost of reputational risk and brand integrity from pollution impacting the local community water supply and arising public health (social and natural capital). It should also include business interuption costs due to lack of access to raw materials (i.e. water) (natural and financial capital), remediation of the natural environment (financial, social, and natural capital) as well as other applicable company, industry and site specific issues.

Concluding this section, the result of an impact pathway valuation for a company is a better understanding of the true value (cost) of its impacts and dependencies. In this ergonomics example, employee and worker injuries and ill health factors into the cost, and ultimate decision, to implement the proposed risk mitigation that will drive a positive impact on human and social capital.

The National Safety Council (NSC) Research

The NSC *New Value of Safety in a Changing World* research report builds upon the impact pathways work introduced by the Capitals Coalition in their natural, human, and social capital protocols. This author participated in the NSC expert group convened for the multi-stakeholder process that yielded groundbreaking research on impact identification, valuation, and value created by the contribution of workplace health and safety in an organization. The research is avaliable on the NSC website. It highlights the interconnectivity of environmental, health, safety, sustainability, and ESG and provides activation guides that begin to assist with integrating these concepts in an organization's business and operations (National Safety Council 2023).

See Table 13.6 for key takeaways from the NSC research report. This research can help companies and individuals identify and articulate a new way of thinking about safety and health, creating value for a company and

Table 13.6 Key takeaways from the national safety council the new value of safety in a changing world research report.

Key takeaways **NSC the new value of safety in a changing world research**
It is important for Safety and health professionals to: (1) be present with our stakeholders. (2) understand and communicate who we are and showing up in a new, different, and material way that impacts a business. (3) translate the math and dollars (your currency) into financial figures to financially demonstrate the value of worker safety and health.

Source: Dony and Seabrook (2023).

its stakeholders as opposed to being a burden to business. In an interview with John Dony, Vice President Workplace Strategy at NSC, an important finding from the research report "is about (the safety and health profession) meeting the world where it is and understanding and embracing the change that's happened." He goes on to say, "the ways safety and health professionals show up, demonstrating the dimensions and interconnectivity of what we do, is an important aspect of value creation for a company" (Dony and Seabrook 2023).

Another important finding from the research is "(safety and health professionals) having the language and the math to demonstrate how safety and health shows up on the balance sheet in a more material way. This has always posed a challenge. There is an intrinsic understanding why people and their health safety and wellbeing is important, but this does not always show up as a material risk on the balance sheet as it has not been easily demonstrated in the past" (Dony and Seabrook 2023). This research provides the methodology, along with activation guide guidance to begin to do the math and demonstrate safety, and health is tied to financial materiality.

The NSC research suggests the correlation between safety and health and financial valuation is still in its infancy. However, the applicability of these tools, specifically the impact pathways/framework, provides a methodology to conduct a valuation of human capital, occupational safety and health, social capital, business relationships, and natural capital as they impact a company and its stakeholders. This valuation is a starting point from which a company can build upon and evolve in its capitals integration and valuation maturity over time (National Safety Council 2023).

In the research report's *Activation Guides*, a case study by Nike is highlighted. A value assessment was conducted for Nike's Supply Chain Safety and Health Initiative. In the assessment, a significant impact identified was the reputational value of the initiative. See Table 13.7 for a breakdown of the monetary value calculated by Nike to its business, value chain's business, and society.

Value Mapping

In the context of sustainability, EHS and human, social, natural, and produced capital, a value map is a visual depiction of a company's impacts and dependencies on people, business relationships, the natural environment, and society and their corresponding value to the company. Tables 13.2 and 13.7 provide the data to develop a partial value map for Natura and Nike.

Value mapping is most effective when monetized units are used as they can be normalized and are universally recognized by business leaders. They are a visual communication tool to facilitate discussions and decision-making, using a common understanding and language, and a consistent, comparable, and relevant valuation process. Value maps can take the form of a table, list, bar, pie chart, heat map, or another graphic image.

Table 13.7 Reputational value of supply chain safety and health initiatives.

Nike Value created by the supply chain safety initiative			
	Nike business value (US$)	**Societal value (US$)**	**Supplier business value (US$)**
Improved quality of life, avoided household costs, Career and Income Opportunities, reduced costs to society, and belonging		170,309	
Business continuity	51,173		
Business reputation	6,650,243		
Cost savings			98,896
Productivity			520,726
Totals	6,711,416	170,309	619,622

Source: Adapted from National Safety Council (2023).

Value mapping should highlight the benefits and value of mitigating impacts and dependencies for all stakeholders, while also communicating a company's positive impacts that provide an advantage over competitors. Internal to a company, value mapping is an important collaborative communication tool informing decision makers on a robust business strategy, including prioritization, internal and external communication, accountability, and resource allocation.

Case Study

First Student is the leading provider of school transportation in North America, operating in 43 U.S. states and nine Canadian provinces. The company provides a good example of aligning on purpose, capability, customer need, and financial performance (Hill and Seabrook 2023). The First Student case study in Table 13.8 demonstrates the value of integration: sustainability, the capitals, the safety and health of passengers and bus drivers, financial performance, customer satisfaction, brand integrity and loyalty and competitive advantage.

Table 13.8 First student case study: aligning on purpose, capability, customer need, and financial performance.

Aligning on purpose, capability, customer need, and financial performance
"Caring for our students and employees is our first priority" says Darryl C. Hill, senior vice president safety and security at First Student. "We complete more than five million journeys each school day, and there is nothing more important than providing a safe environment on our buses for our students' and drivers' safety." First Student is the leading provider of school transportation in North America, operating in 43 U.S. states and nine Canadian provinces. Capability, services, trust, reliability, and partnerships are key to its longer-term strategy for success (Hill and Seabrook 2023).
First Student serves more than 1300 customers, primarily school districts, whose families can benefit from its experience and expertise. It is a privately held company, which means it also has a fiduciary duty to meet stakeholders' expectations, including financial performance. To exceed customer expectations, and create value for all its stakeholders, First Student focuses on innovation, sustainable practices, and operational excellence, helping to ensure school buses remain the safest way for students to get to and from school.

(*continued*)

Table 13.8 (Continued)

"Broader societal impact of our company's activities include providing student passenger care to meet customer expectations," according to Hill. Student passenger care builds trust and confidence among parents, communities, school boards, and the students themselves. Hill shares, "Drivers manage challenging situations and behaviors among all passengers on their bus." This is why First Student has partnered with Cincinnati Children's Hospital, one of the top children's hospitals in the United States to create the FirstServes™ program (Hill and Seabrook 2023).
FirstServes provides education and support to school systems, parents, and caregivers in order to facilitate the safety and well-being of student passengers (First Student 2024). The program helps create a safe and secure journey for student passengers and drivers. One-third of First Student's routes accommodate special needs students, and FirstServes ensures the company provides a support system for all students (Hill and Seabrook 2023).
According to customer feedback from one school district, First Student "bridged the gap between the classroom and the school bus by implementing FirstServes. The collaboration resulted in improved employee satisfaction and student experience, with fewer student incidents on the bus" (First Student 2024).
FirstServes also empowers drivers with the skills to support students, creating opportunities for individualized care. The program was implemented in select school districts during the 2022–2023 school year and immediately resulted in positive behavioral outcomes. Company data showed a 27% decrease in behavioral incidents on buses. "Drivers feel a sense of purpose and responsibility knowing a good student experience begins and ends on the school bus," according to Hill. That sense of purpose and a driver's alignment with First Student's overarching purpose and commitment to safe student passenger journeys engenders driver loyalty. The outcome is committed drivers and high driver retention rates (Hill and Seabrook 2023). Aligning on purpose and the resulting loyalty are hallmarks of good human capital management, as well as the social or people side of sustainability and ESG.
School districts' transportation needs continue to evolve, and over the past several years, First Student has seen a trend of more districts prioritizing comprehensive solutions over pricing when awarding bussing contracts (Hill and Seabrook 2023). Hill sees First Student as a full-service transportation partner, leading with care, technology, and innovation. The results are proprietary products and services that provide an exceptional student experience, setting it apart from any company in the industry (Hill and Seabrook 2023).

Source: Hill and Seabrook (2023)/with permission of Darryl C. Hill.

This case study brings home Accounting for Sustainability Executive Director, Helen Slinger's view that "the ultimate goal is sustainable business being business as usual. . ." (Slinger and Seabrook 2023).

Summary

This chapter has introduced the concepts value creation, value assessment, valuation, and impact pathways. By providing examples of impact pathway value assessments and outputs, it should enable companies to create bespoke valuation scenarios, engaging a cross section of internal and external stakeholders and to capture monetized value in the profit and loss reports. In this chapter, we have used the Capitals Coalition impact pathway valuation process and the National Safety Council (NSC) The New Value of Safety in a Changing World research as a starting point. These value assessment processes are adapted from the Capital Coalition, International Foundation for Valuing Impact and Value Balancing Alliance impact pathway valuation methodology. Research presented from the NSC activation guide examples are depicted in monetary terms that can have a direct influence on a company and its stakeholder's decision-making.

References

Bauer, A. and Seabrook, K.A. (3 October 2023). Interview with Angus Bauer, Head of European Equity Research, Schroders. Bauer, Angus (interviewee) and Seabrook, Kathy A. (interviewer).

Capitals Coalition (2016). Natural capital protocol. http://www.naturalcapitalcoalition.org/protocol (accessed 11 October 2023).

Capitals Coalition (2019). Social and human capital protocol. https://capitalscoalition.org/capitals-approach/social-human-capital-protocol (accessed 8 August 2023).

Capitals Coalition (2021). Social and human capital protocol: A primer for Business. https://capitalscoalition.org/wp-content/uploads/2021/12/Social-Human-Capital-Protocol-A-Primer-for-Business.pdf (accessed 17 August 2023).

Capitals Coalition (2022a). Valuing nature and people to inform business decision-making. Coursera Online Course. https://capitalscoalition.org/new-online-business-training-course-launched-today (accessed 23 September 2023).

Capitals Coalition (2022b). Valuing human capital in occupational health and safety. https://capitalscoalition.org/wp-content/uploads/2022/11/Valuing-Human-Capital-in-Occupational-Health-and-Safety.pdf (accessed 8 January 2024).

Dony, J. and Seabrook, K.A. (2023). 23 August 2023 Interview with John Dony, Vice President Workplace Strategy, NSC. Dony, John (interviewee) and Seabrook, Kathy A. (interviewer).

First Student (2024). FirstServes™ program. https://firststudentinc.com/innovation/first-serves (accessed 9 April 2024).

Global Solutions, Inc. and Seabrook, K.A. (2024). 6 August 2024 seminar on ESG/sustainability: strategy, road mapping and action planning. American Society of Safety Professionals, preconference seminars, Safety 2024.

Hill, D.C. and Seabrook, K.A. (2023). 25 September 2023 Interview with Darryl C. Hill, Vice President safety and security, first student. Hill, Darryl C. (interviewee) and Seabrook, Kathy A. (interviewer).

Impact Management Platform (2024). Impact and the impact pathway. https://impactmanagementplatform.org/impact (accessed 19 May 2024).

International Foundation for Valuing Impact and Value Balancing Alliance (2023). General methodology I: conceptual framework for impact accounting (exposure draft). https://www.value-balancing.com/_Resources/Persistent/5/0/b/0/50b01884def43e1cca7ac47d2544dc3ef9daea32/IFVI_VBA_Public%20Exposure%20DRAFT_General%20Methodology%201_A4.pdf (accessed 18 May 2024).

National Safety Council (NSC) (2023). The new value of safety and health in a changing world. https://www.nsc.org/getmedia/b96b7c20-994c-4959-9a36-a8507a2be7e7/nvos-report.pdf (accessed 23 September 2023).

Natura & Co (2022). Natura integrated profit & loss accounting 2021 - technical executive summary and insights. https://api.mziq.com/mzfilemanager/v2/d/9e61d5ff-4641-4ec3-97a5-3595f938bb75/d8f2cae6-7a86-1d24-8100-62ae5871c7fc?origin=2 (accessed 2 January 2024).

OECD (2024). The OECD Framework for Measuring Well-Being and Progress. Measuring Well-being and Progress: Well-Being Research. https://www.oecd.org/wise/measuring-well-being-and-progress.htm (accessed 15 February 2024).

Slinger, H. and Seabrook, K.A. (2023). 25 September 2023 Interview with Helen Slinger, Executive Direct, Accounting for Sustainability. Slinger, Helen (interviewee) and Seabrook, Kathy A. (interviewer).

Stangis, D. and Seabrook, K.A. (2023). 18 July 2023 Interview with Dave Stangis, Chief Sustainability Officer. Stangis, Dave (interviewee) and Seabrook, Kathy A. (interviewer).

Stangis, D. and Smith, K.V. (2017). *The Executive's Guide to 21st Century Corporate Citizenship*. Bingley, UK: Emerald Publishing Limited.

Sustainable Business Now (2024). How to integrate sustainability into finance and decision-making: natura's integrated profit & loss. https://www.sustainablebusinessnow.org/posts/natura-integrated-profit-and-loss (accessed 29 June 2024).

World Health Organization (2020). WHO methods and data sources for global burden of disease estimates 2000–2019 https://cdn.who.int/media/docs/default-source/gho-documents/global-health-estimates/ghe2019_daly-methods.pdf?sfvrsn=31b25009_7 (accessed 30 June 2024).

14

Disclosures and Reporting

Introduction

Note to the reader: As you will recall from Chapter 5 *Frameworks*, there are numerous regulatory and nonregulatory (or voluntary) frameworks, both publicly available and proprietarily protected. This chapter will touch on a number of those frameworks as examples of a specific reporting discussion, but will not cover each in detail; therefore, the reader should refer back to the Frameworks as you explore sustainability disclosures and reporting. In this chapter "framework" is intended to represent the collective body of principles, guidelines, voluntary standards, and regulatory reporting requirements.

> Fun Fact – Denmark was the first country to require disclosure of environmental reporting in the mid-1990s as part of financial reports (European Environment 2005).

As introduced in Chapter 2 *Evolution of Sustainability*, the "business of sustainability" including the disclosures and reporting on sustainability performance outcomes, has increased a hundredfold since 2002 (Harvard Business Review 2021). As per Peter Drucker proclaimed "what gets measured gets managed." However, absent widespread regulatory requirements, organizations are still driven to sustainability disclosures. Reporting enhances performance, improves corporate interest rates, increases investment and earnings, and to drives toward more aligned reporting within their peer group or industry (Harvard Business Review 2021). The number of organizations formally publishing a sustainability report or incorporating sustainability into their financial reporting has continued to increase with over 90% of the world's top companies reporting on sustainability in 2022 (Journal of Accountancy 2024).

Connecting the Dots: Environment, Health, Safety, and Sustainability,
First Edition. Kathy A. Seabrook and Tanis J. Marquette.

Organizations should consider reporting requirements early in their sustainability journey – not because there are so many reporting requirements (although that is true), but more so because if how to communicate material issues is not a consideration, it becomes a challenge at reporting time, as the authors have experienced. As the old adage goes, garbage in, garbage out. And without good data collected for internal stakeholders, understanding what is truly driving sustainability may remain unknown or influence poor practices (Forensic Risk Alliance 2024; Markersite 2024). This Chapter provides an overview of reporting and disclosures in an attempt to avoid having your organization live the old garbage in-garbage out adage.

What Is It?

To ground the Chapter, lets first define the similarities and differences between reporting and disclosures since the words are used seemingly interchangeably (Wikipedia 2024). From Investopedia, we learn that disclosures are simply the "process of making facts or information known to the public" and may appear in fine print in a 10-k (Investopedia 2024). In fact, the United States SEC requires disclosure statements in all financial research reports (US SEC 2024). That is, the SEC requires the reports to include statements disclosing the relationship between the report author and the issuer company (which is the subject of the report). Whereas in sustainability reporting, the report is generally more about sharing status on performance related to nonfinancial risks and opportunities (Ecovadis 2024).

> For a holistic introduction to global disclosures and reporting, both voluntary and mandatory, the website Carrots and Sticks (www.carrotsandsticks.net) maintains a listing of many current and pending frameworks. The site allows the user to sort by voluntary or mandatory, issuer type, topic, region, or sector. Further, the site also provides policies in 40 languages, providing greater access to non-English resources. Carrot and Sticks is a joint effort by GRI, King's College London, The University of Edinburgh, and the Stellenbosch Business School. Periodic reports posted on their website provide an analysis of global reporting. The website provides information on disclosure requirements, reporting frameworks, sustainability policy initiatives, and other resources).

Sustainability reporting and disclosures have continued to evolve, stemming initially environmental regulatory reporting to primarily voluntary sustainability reports; recent evolution has focused nonfinancial reporting toward more robust and industry-aligned mandatory reporting, like the EU's Corporate Sustainability Reporting Directive (CSRD) and Corporate Sustainability Due Diligence Directive (CS3D) (Wikipedia 2024). Beyond regulatory frameworks, as previously noted, others can be voluntary. In fact, you may also recall from Chapter 2 *Evolution of Sustainability* that the International Financial Reporting Standards Board, which oversees financial reporting requirements for hundreds of countries, has issued two sustainability reporting standards since absorbing SASB and TCFD. Both S1, the General Requirements for Disclosure of Sustainability-Related Financial Information, and S2, the Climate-Related Disclosure Standard, are currently voluntary, but have been officially adopted by 16 countries with Brazil announcing it will become mandatory in 2026 (ESG Today 2023).

The IFRS S1 and S2 ISSB Sustainability Standards may remain voluntary even if absorbed or referenced into regulatory requirements, if revenue thresholds referenced in the implementation plans of said jurisdictions have not been met by companies within those jurisdictions or if the company's employee or trade profile does not meet the regulatory description. Another such example may be the Global Reporting Initiative, now known only as their acronym GRI Standards, which are at the time of this writing, voluntary, but under consideration for reference by some EU regulatory development bodies through an instrument called a Memorandum of Understanding (MOU). The MOU outlines how each independent framework (which could be any of the examples we have discussed (guidance, voluntary standards, and/or regulatory standards) works together to achieve a stated goal or outcome.

Across sustainability, whether reporting and disclosure frameworks, ESG Raters, or for-profit sustainability assessment services, different bodies are merging, being acquired, or otherwise engaging to address the confusion of the landscape. However, in spite of all the sustainability market consolidation, even within the EU's jurisdiction, GRI has remained the leading (voluntary) reporting framework since 1999. In fact, GRIs guidance documents, and now standards, are wholly focused on encouraging company double materiality disclosures. Many sustainability reports will indicate the report is aligned with GRI Standards or include an index referencing specific GRI disclosures (Figbytes 2024). The GRI Standards promote a consistent, organized, and structured approach to sustainability reporting that benefits not only the reporters, but readers alike (National Center for Corporate Reporting 2024).

> According to the National Center for Corporate Reporting, using the GRI Framework Standards:
>
> - can be used to benchmark organizational performance with respect to laws, norms, codes, performance standards and voluntary initiatives,
> - demonstrate organizational commitment to sustainable development, and
> - compare organizational performance over time.

However, not all "mandatory reporting" stems from regulation. A voluntary framework that has been adopted by an organization as their reporting platform of choice can, by virtue of the business relationship, create the "impression of a mandatory reporting framework." By having requested that suppliers use a specific reporting framework, like a third-party assessment, for example, and due to the business relationship between the requesting and reporting organizations, while the framework cited is not regulatory, it can become "mandatory" and may be included in contracts or other business instruments. For example, organizations that belong to Together for Sustainability (TfS) have a membership or contractual requirement to request their suppliers' complete questionnaires and have their documentation substantiated by Ecovadis or another approved audit vendor (Together for Sustainability 2024). While it is not a mandatory regulatory requirement to complete an Ecovadis assessment, the reporting organization may consider the assessment "like a regulatory requirement" due to underlying business-to-business agreements.

Together for Sustainability (TfS) is an industry-led effort to provide greater transparency into chemical product supply chains (Together for Sustainability (2024). The initiative has a decision-making body of 50 members representing organizations from across the chemical industry value chain. The member companies provide assessments to their business partners to evaluate the potential for sustainability risk in the supply chain. The assessments, which may be an online questionnaire or an onsite audit, gauge the supplier's sustainability performance toward ultimately improving that performance. Information gathered in the assessments is shared across the membership to reduce redundancy in reporting.

Ecovadis, a partner to TfS, is a supply chain assessment company. Ecovadis provides proprietary sustainability questionnaires developed from widely accepted frameworks, namely, the Global Reporting Initiative (GRI), the United Nations Global Compact, Responsible Care®, and ISO 26000 Guidelines for Social Responsibility. Companies submit their completed questionnaires and supporting documentation toward achieving Ecovadis

Medals. Whether as part of TfS or a separate Ecovadis client initiative, completed scoring can be shared in some circumstances with other Ecovadis clients also with the goal of reducing reporting (Together for Sustainability 2024). Both TfS and Ecovadis use their influence as quasi-mandatory reporting frameworks to increase reporting within the supply chain.

With so many reporting frameworks and disclosure requests, it is no wonder professionals and organizations alike are overwhelmed when considering how to approach reporting. How does an organization fit its materiality into the frameworks? What performance data is necessary? What do you do about gaps? Do the gaps even matter? Does submitting to Ecovadis really reduce reporting burden? It has been reported that companies in the EU may be investing as much as 500,000 euros in sustainability reporting (Ecovadis 2024)! To help answer these questions, let us step back and consider the who, what, when, where, why, and how of sustainability reporting.

Sustainability Reporting Decisions

In many organizations, the authors' experience sustainability reporting begins months before the inaugural report and reporting for the *second* report usually starts about that same time! Let us explain. . .

As organizations work through their data-story continuum they are answering critical questions about their audience and reporting rationale within their available resources (Conservice ESG 2024). The first question that may go unasked, is why are they reporting in the first place. The answer may be lost in the efforts to gather all the necessary resources and data, or perhaps it is not clearly communicated throughout the organization. However, back at the design offices, the *why* helps to answer the *what*. The materiality assessment, when conducted, may answer the *how* of the what and for *whom*. The *when* and *where* come later in the process. Wait. . . . **What?!! Confusing, right?**

Let us start with **Why** an organization would publish sustainability disclosures in a report. **Why** the organization has chosen this time to begin reporting? In the authors' experience, organizations report for a combination of reasons. Some more prevalent reasons to begin reporting include:

- *Peer Pressure* – organizations in their peer group are publicly reporting their achievements and the organization does not want to be seen as a laggard
- *Customer Demand* – either through contract or request, customers may be asking for certain disclosures and with all that data amassed, the organization chooses to publicly report and gain benefit from the reporting and sharing their stories

- *Regulatory Requirements* – while some regulatory requirements underly the reporting, augmenting with a fuller sustainability report allows the organization to tell their story their way
- *Materiality* – stakeholder feedback would indicate the organization's stakeholders want to understand organizational sustainability performance

There are certainly many more reasons organizations report on their sustainability performance. See Figure 14.1 The who, what, when, where, how, and why of sustainability reporting (Marquette 2024).

Why the company has chosen this time to begin reporting, may give rise to what to report. Peer reporting may be included in the benchmark assessments we discussed in Chapter 11: Benchmarking. Reviewing what is included, and not included, in the public sustainability reports and websites of similar companies in the industry or companies of similar revenues may yield insight into material issues, disclosure frameworks embraced, goals, and initiatives. The peer benchmark reporting considerations may also identify areas in your organization that are leading or lagging in the market. Customer feedback may also provide a unique window into material issues for the value chain; issues and impacts the stakeholders want to understand about your organization.

As discussed in the Materiality chapter, materiality assessments also offer insights into **What** holds stakeholder interests, can inform strategy, and/or provide the foundation for reporting. If your materiality assessment considered customers as a unique stakeholder group in the analysis, the organization may be able to combine materiality, disclosures, and messaging to further the commercial marketing strategy, for example.

As regulatory requirements have increased in scope and coverage, they have also become a possible index for reporting, just like the GRI standards. In fact, there are now spreadsheets that cross-reference regulatory and voluntary standards – as they say, "just Google it." The interoperability references were intended to show alignment between the frameworks but could also be used to start the conversation on what to include in a report. GRI has published on their website the GRI disclosure standards comparative between each version. Further, with the introduction of the EFRAG ESRS, you can access entire Microsoft © Excel workbooks of the ESRS requirements (although given the extensive nature of the metrics and disclosures, the authors would recommend only selecting relevant line items (i.e., material sustainability matters) else the list will be undoubtedly overwhelming)

> The authors have learned, through engaging the ESG Raters, if striving to increase ESG ratings, sustainability reporting electronic files (pdfs) should parallel website stories to ensure ratings companies have the opportunity to capture keywords during web crawler searches.

WHY report?	WHO is the audience?	WHAT framework to use?	WHEN to report?	WHERE to publish?	HOW to get it done!
Peer pressure – organizations in their peer group are publicly reporting their achievements and the organization does not want to be seen as a laggard	Even sustainability reporting initiated through the regulatory process may have an audience well beyond the regulatory authority.	With no regulatory or customer defined framework required, the organization may loosely follow any number of recognized frameworks. However, best practice would be to align with the organization's industry standards, whether through GRI or SASB	Reports written to meet the organization's own strategy or to increase reputation may be issued on the organization's arbitrary timeline or to coincide with Board meetings, for example, for ease of reviewing.	Organizations will provide their disclosures, as required, to specific frameworks like Ecovadis and CDP, but may often publish these unique disclosures on their website. In some cases, the reporting creates the outline for the organization's sustainability webpages.	Understanding the who, what, when, where, and why of sustainability reporting and disclosures together with where to locate internal sustainability champions, data-owners, and storytellers helps to begin to answer the How. How.... can this effort be successfully completed with the resources we have internally or do we need external support to help us meet the timeline?
Customer demand – either through contract or request, customers may be asking for certain disclosures and with all that data amassed the organization chooses to publicly report and gain benefit from the reporting	Specific framework disclosures may be prepared for that entity but shared publicly with all stakeholders. One example of this broader disclosure sharing is CDP submissions that can be found on many company websites.	Customers and suppliers will request their value chain partners use specific frameworks or provide certain disclosures. Examples are Ecovadis, CDP, and TCFD and may even include issue-specific frameworks like SBTi.	Value chain reporting may be driven by framework schedules (e.g. CDP submitters all meet a July deadline) or other industry-specific schedule.		

Figure 14.1 The who, what, when, why, and how of sustainability reporting.

WHY report?	WHO is the audience?	WHAT framework to use?	WHEN to report?	WHERE to publish?	HOW to get it done!
Regulatory requirements – while some regulatory requirements underly the reporting, augmenting with a fuller sustainability report allows the organization to tell their story	While materiality assessments inform the organizations key issues and specific disclosure drivers, materiality assessments are also often included in sustainability reports for all audiences	Regulatory reporting, in addition to generating some of the underlying data, may also provide the rough outline for sustainability reporting, if not the entire report (as in the case of the EU CSRD).	Initial reporting will align to the regulatory requirement, but may be augmented with update reports later in the year. Many US-based companies include material ESG issues in their 10-k and publish a sustainability report later in the reporting year.		Is the organization resource or time constrained?
Materiality – stakeholder feedback would indicate the organization's stakeholders want to understand organizational sustainability performance		Stakeholder feedback, in particular from investors, may also provide guidance as to what to report. Investors may rely on answers to ESG Rater surveys, like S&P Globals SAM or Sustainalytics' issuer questionnaire.	Stakeholder-driven reported will be based on the stakeholder group, however, ESG Rater and Investor reporting may coincide with regulatory financial reporting and the ESG Rater Scoring schedules.		

Figure 14.1 (Continued).

We have covered the why and what, which also provides insights as to the "**Who**" the organization is considering as its reporting audience. The report could be focused on the customers, regulators, or any combination of the broader stakeholder groups. It may also include the ESG rating agencies. The ESG rating agencies review and score the publicly traded company's financial statements, sustainability website information, and released reports. If preparing a report for this audience, it is helpful to engage with the organization's board of directors to understand which rating agency reports they read. From our earlier discussion on rating agencies, we know that each has their own proprietary questionnaire and weighting from which to build the ESG risk profile or score. If the disclosure strategy includes a goal to increase (or decrease as the case may be) rating agency scoring, noting the agency's key issues and aligning data disclosures and stories may prove helpful.

When to release the report may coincide with industry announcements, internal organizational goals, regulatory compliance dates, or a publicly traded organization may want to release their report to provide sufficient advance notice to the ESG ratings companies to (ideally) receive an improved score in their next scoring update. The ESG rater scores can influence board member and investor decisions and financial transactions, including interest rates and credit scoring (Simply Sustainable 2024). The timing may also be governed by the Board or leadership meeting schedule. This is particularly true when a Board does not have a specific sustainability committee and sustainability oversight is incorporated into the main Board meeting. The sustainability committee or Board usually has the opportunity to provide feedback on the draft report before publication.

Where to disclose or publish the written report? If the only why is because the customers or investors have requested a CDP submission, for example, submitting and allowing disclosure of the submission through the online system may be all that is initially required. Over the next reporting period, the organization can work behind the scenes to better understand the value of external reporting, review their benchmark information and materiality assessment, and then more purposely publish in the next year. Other organizations may use the inaugural report to enhance their website and make their website page updates.

However, if the inaugural effort is more substantial, the initial report may be launched on the website with a news release over the AP wire, an email campaign (email blast) to customers and vendors (business partners), and updates to the website. While rarer, organizations with larger offices that still host meetings and in-person events will likely print copies of their report and accompanying marketing materials. If the organization is vested in their sustainability success, the marketing team may launch a marketing campaign around sustainability reporting and update the website on a quarterly basis to keep stories fresh and evergreen. This keeps the website renewed, keeps employees engaged, and keeps people interested in sharing their stories. This stakeholder engagement approach enables cross-functional teams to embed sustainability across the organization and build resilience into processes and practices as discussed earlier in Section I.

This leaves the **How**. The how will be, no doubt, unique to each organization, but us having shared a few lessons we have learned along the way will make your how a little smoother. Regardless of where your organization starts on the reporting journey, as you gather and begin to analyze their footprint data and marry that data to the stories you hope to tell, a realization occurs. What if we had different data? Could we tell the story differently? Or, this was the story we thought we were going to tell, and then, we looked at the data! The authors have found that the sustainability reporting process as much defines the current report as it does the future report and, in fact, the strategy overall.

However, we would be remiss if we did not draw your attention, if just for a moment to greenwashing and once there, to bluewashing. But first greenwashing. Greenwashing is a term that has been tossed about since the late 1980s (Climate Club 2024). It was coined by a man named Jay Westerveld, referencing a hotel saving hotel laundering costs by not washing towels to save the environment. Since that time, the US Federal Trade Commission (US FTC), the United Kingdom Advertisers Standards Authority (UK ASA), and Australian Competition and Consumer Commission (ACCC) are just a few of the global regulatory bodies that have issued greenwashing guidelines or standards. Each of these standard bodies set out requirements in advertising and marketing for appropriately substantiating environmental claims directed at consumers, and in some jurisdictions business-to-business marketing. Substantiation may include maintenance of documentation records and appropriate calculations, in particular for product-related statements.

As your sustainability teams and authors are collecting data and writing your organizations' sustainability report, be sure to review appropriate jurisdictional marketing standards. Environmental engineers in our readership may also be familiar with the parallel requirements of ISO Product Ecolabeling Standards (ISO-14020-14027).

Bluewashing is an emerging term referring to sustainability communications presented by companies that have signed the United Nation's Global Compact but have yet to be seen as making significant policy changes that will impact change (Forbes 2022; Sarvahitey 2018). Blue washing, or social washing, gets its name from the blue color of the United Nations flag. Also evolving as biodiversity has become commonplace in sustainability reporting, bluewashing may also represent misrepresentation of environmental claims specific to marine life or water stewardship. However, most references to bluewashing relate to communications regarding an organization's UN pledge or commitments (See2See 2024).

With changing regulation and the dawn of ESG, the financial institutions and processes, including auditors and assurance departments began to embrace the spectrum responsibilities of ESG. Accountants began embracing their responsibility for oversight of sustainability data. A common misconception the authors have experienced is that the environmental health and safety professional "owns" the greenhouse gas data. While in some organizations, the EHS team may be responsible for those calculations, the actual data is generated from the utility invoices stemming from operational equipment and utilities, and therefore "owned" by operations, but often resides in finance. The EHS team is "borrowing" the data to perform the calculations and report upward or utilize the data for projects and regulatory or permit reporting. The EHS team must work in collaboration with operations to initiate any improvements or modifications to process or equipment to reduce emissions, which are often tracked by finance (or Six Sigma teams). Rarely can EHS negotiate rate changes to the energy contracts and ultimately impact utility invoices. It is a collaborative effort.

Likewise, reducing emissions in the supply chain, is a collaboration between procurement and sustainability or EHS or ideally all three teams. Often, the EHS team has responsibility for supply chain on-site audits and lead waste management site inspections and data management. It is a natural partnership to support the sustainability audit while gathering critical EHS information.

As previously discussed in the context of reporting and disclosures, software platforms can be selected to assist your organization through the annual reporting grind, and more.

Todd Fein, CEO of Green Diamond, LLC, a sustainability subject-matter expert and solutions provider, with decades of experience developing and implementing programs in collaboration with Fortune 50 companies, discusses how to select the "right" platform.

Todd suggests "The right platform should be a force multiplier. It should not only be capable of handling the annual reporting grind, but also offer the additional features and functions necessary to help organizations analyze, identify, and pursue opportunities to decrease risks and/or costs and increase goodwill and/or sales the other 364 days of the year. Ideally, it should also come from a partner with sustainability or ESG expertise the organization can tap, not just a software or platform shop. In other words, the right platform should meet you where you are and want to go on governance, materiality, maturity, innovation, and (strategy) change management as much as deliver digital enablers like integration, automation, visualization, simulation, and one version of the truth" (Fein and Marquette 2024).

He continues, "Of course, annual regulatory compliance and external reporting are crucial, but they are also just the table stakes. The kind of success regulators, investors, and other stakeholders want is for companies to actually achieve performance that's sustainable. So, avoid platforms that focus on just one or two dimensions. The platforms purpose-built to cover the full breadth of sustainability and ESG (economic prosperity, environmental stewardship, social responsibility, corporate governance) that illuminate how financial and non-financial impact metrics correlate and cause one another to improve or worsen, that offer predictive analytics and prescriptive improvement actions, that scale and flex with the business in the front and back offices. . . those are the gems. A colleague and former head of global sustainability for a Fortune 50 and I sometimes like to use the metaphor of a sports game to differentiate platforms only capable of helping with the annual reporting grind and those capable of also enhancing decision-making in operations and innovation. We say the first kind help organizations manage the 'scoreboard,' the second kind help them manage the 'field,' and we all know where the scoring happens, and the game is either won or lost" (Fein and Marquette 2024).

Todd finishes with a reminder, "Systems are not solutions, and they are only as valuable as their adoption. You don't want to have to switch platforms later, so you should refuse to compromise on having anything but a platform that's robust enough to go broad, deep, and work 365 days a year for you as you mature and sustainability becomes more embedded, business as usual. Of those, however, you should try to discern the ones that are more modular, more configurable, and offer affordable, easy setup options, and even offer a free proof of concept. It's hard to overestimate the value of a robust platform that can still keep it simple to start, let alone a vendor willing to put some skin in the game to show they are a purpose-driven partner with a long view, not just a profit-making machine too short-sighted to take care of the client relationship" (Fein and Marquette 2024). Sage advice for those embarking on adventures in software procurement, no doubt!

Who Is Generally Involved?

Whether wholly developed in-house or through an external market communications team, sustainability reporting historically had no specific release date. Depending on the data-gathering process and whether the organization used direct invoice data or data estimates, some reports were not published until late Spring, early Summer. However, the newly released EU CSRD regulatory reporting requires sustainability reporting with the annual financial report; the authors anticipate, as we have seen with the adoption of health and safety rules organizations reporting in alignment with the EU Directive will likely align their sustainability report coincide with their annual reporting, even in other jurisdictions to reduce reporting burdens.

The foundation of sustainability reporting is the data generated by various functions across the organization. In a manufacturing organization, many of the performance datasets are generated from operations or production data. Products may also be at the center of a service organization's data. (Note: The Greenhouse Gas (GHG) Protocol provides specific guidance on defining the scope and inventory of greenhouse gases for inclusion in an organization's GHG management plan, including how to estimate and calculate emissions. The Protocol can be found at https://ghgprotocol.org.)

The process of capturing, collecting, collating, and analyzing the sustainability performance data can be assigned to a number of roles or teams within the organization. The process may be defined based on the organizational hierarchy, the data owners, or sometimes the system of record, as Todd Fein from Green Diamond mentioned. The data management process may initially be seen as an initiative, in particular when the organization is publishing its inaugural sustainability report. The authors strongly encourage organizations to develop a consistent and repeatable process to manage data. Repeatability of a defined process reduces potential for error and allows your organization to capitalize on the opportunity for engagement offered through the report development process.

Once the data is "managed," report-authors can engage with employees and other stakeholders to tell "their story" within the organization's sustainability story. In the authors' experience, internal stakeholders will be more supportive of the sustainability journey if they can see themselves within the story and relate to experiences.

The full sustainability report comes together as the authors tie the data, analysis, and the stories together with leadership's vision, goals, targets, and future planning. The report may share updates on programs, realignment of goals, initiatives undertaken to achieve goals, and awards and honors received throughout the year or since the last report. The governance section may include updates on annual training, the leadership team, and the Board of Directors. There may also be a policy section with links to current policies and procedures, if the report is available online.

In our earlier chapters, we discussed stakeholders and materiality. Most sustainability reports will include not only the materiality matrix but also a narrative describing the materiality assessment process. It may include the stakeholder engagement process and a table of stakeholder engagement strategies. However, each report, like each organization, is unique, and not only are no two alike, but no two are alike from year to year, even from the same organization. If you are conducting a benchmark assessment of a peer company, be sure to review sustainability reports going back a few years, however. The authors' experience has found that some organizations may report on materiality only periodically, which means if you happen to select an "off" year, you may not find all the sections, like a materiality matrix, you are looking to find in each edition.

Where Is It Interconnected?

Sustainability disclosures and reporting are bound by a thread that weaves throughout the organization. While it may give the appearance that reporting is at the end of the cycle, it is really a continual loop as we saw in the Stakeholder Influence graphic as discussed in Chapter 10: Stakeholders and Stakeholder Mapping Organizations are continually planning their updates and stories, goals and targets, and initiatives. As you research organizations, most companies have webpages dedicated to their sustainability journeys and many will use that landing page to update their activities between formal reporting cycles.

Why It Has Value?

- How it fits into strategies and activities

Sustainability reporting provides the organization the opportunity to tell their story their way. While the regulatory frameworks have streamlined the content, they have also added another element to reporting that was previously reserved for financial reporting – assurance. That is not to say there are no examples of historical sustainability reporting that has been assured; however, with the consolidation of the frameworks and the introduction of the EU's Corporate Sustainability Reporting Directive (CSRD), limited assurance is required initially and moves to reasonable assurance in the future (PwC 2023). By definition, reasonable assurance is reasonably stated and materially corrected, whereas limited assurance is a lesser level of assurance and not as comprehensive a review. Both are performed by a third party that issues an assurance letter.

The sustainability report may be reviewed as part of an ESG ratings review, one of the stakeholders interested in an organization's sustainability performance on behalf of investors. The report may also be the first part of a larger engagement with stakeholders overall, which leads us to the benefits of reporting. According to Workiva (a leading platform for integrated reporting), sustainability reporting assists an organization to:

- Build trust with consumers
- Appeal to ESG investors
- Create more efficient internal processes
- Differentiate your business
- Make meaningful, measurable climate progress

However, from GRI's *The Value of Sustainability Reporting and the GRI Standards,* we learn that "much of the strategic value of sustainability comes from continuously talking to and learning from key stakeholders". And, that truly is the long and short of sustainability reporting – it is all about the stakeholders.

As the regulatory landscape continues to mature, sustainability reporting will continue to evolve. Organizations globally will continue to share their stories of their journey on the path to 1.5°C.

References

Climate Club (2024). https://www.theclimateclub.co/sustainabilityblog/the-history-of-greenwashing-and-its-modern-evolution (accessed 11 November 2024).

Conservice ESG (2024). The ESG reporting matrix. https://esg.conservice.com/esg-solutions/the-esg-reporting-matrix (accessed 17 February 2024).

Ecovadis (2024). Sustainability reporting. https://ecovadis.com/glossary/sustainability-reporting (accessed 17 February 2024).

ESG Today (2023). Brazil to require mandatory sustainability reporting from 2026. https://www.esgtoday.com/brazil-to-require-mandatory-sustainability-reporting-from-2026 (accessed 17 February 2024).

European Environment (2005). A decade of mandatory environmental reporting in Denmark. *European Environment* 15 (6): 362–373. https://doi.org/10.1002/eet.397.

Fein, T. and Marquette, T.J. (2024). Interview with Todd Fein, CEO Green Diamond, LLC. Fein, Todd (interviewee) and Marquette, Tanis J. (interviewer).

Figbytes (2024). GRI framework: what is it & what is it important? https://figbytes.com/blog/gri-framework-what-is-it-why-is-it-important (accessed 20 June 2024).

Forbes (2022). Bluewashing joins greenwashing as the new corporates whitewashing. https://www.forbes.com/sites/timothyjmcclimon/2022/10/03/bluewashing-joins-greenwashing-as-the-new-corporate-whitewashing/3 October 2022 (accessed 20 June 2024).

Forensic Risk Alliance (2024). How can data support your sustainability strategy? https://www.forensicrisk.com/news-and-insights/how-can-data-support-your-sustainability-strategy (accessed 17 February 2024).

Harvard Business Review (2021). Overselling sustainability reporting. https://hbr.org/2021/05/overselling-sustainability-reporting (accessed 17 February 2024).

Investopedia (2024). What is a disclosure? In business and how the work. https://www.investopedia.com/articles/analyst/073002.asp (accessed 17 February 2024).

Journal of Accountancy (2024). Sustainability reporting assurance rates of the rise globally. https://www.journalofaccountancy.com/news/2024/feb/sustainability-reporting-assurance-rates-on-the-rise-globally.html (accessed 17 February 2024).

Markersite (2024). Overcoming the hurdles why people processes and Bad Dat are hindering sustainability. https://makersite.io/insights/overcoming-the-hurdles-why-people-processes-and-bad-data-are-hindering-sustainability (accessed 20 June 2024).

Marquette, T. (2024). The kaleidoscope of sustainability. PowerPoint Presentation, Health and Safety Professionals Canada 75th Annual Conference 2024, Edmonton AB, July 13, 2024.

National Center for Corporate Reporting (2024). Benefits of GRI reporting. https://nccr.id/gri-certified-training/benefits-of-gri-reporting (accessed 20 June 2024).

PwC (2023) https://viewpoint.pwc.com/dt/us/en/pwc/news-and-alerts/landing-page/csrdplacematesglp.pdf (accessed 11 November 2024).

Sarvahitey (2018). Working as one for the good of us all. https://sarvahitey.wordpress.com/2018/05/16/greenwashing-and-blue-washing/ (accessed 17 February 2024).

See2See (2024). Beyond the blue facade: unmasking bluewashing and the environmental impacts. https://www.sea2see.org/blogs/sustainable-eyewear-made-out-of-recycled-marine-plastic/beyond-the-blue-facade-unmasking-bluewashing-and-its-environmental-impact (accessed 2 June 2024).

Simply Sustainable (2024). https://simplysustainable.com/insights/the-current-state-of-esg-ratings (accessed 11 November 2024).

Together for Sustainability (2024). http://www.tfs-initiative.com (accessed 17 February 2024).

United States Securities and Exchange Commission (2024). S-K disclosure requirements. https://www.sec.gov/files/reg-sk-disclosure-requirements-review.pdf (accessed 17 February 2024).

Wikipedia (2024). Sustainability reporting. https://en.wikipedia.org/wiki/Sustainability_reporting (accessed 17 February 2024).

15

Resilience

Resilience is a term with many uses and as many professional applications. In this chapter, we are exploring the interconnectedness of a health and safety mindset and the sustainability professional's challenge – that is, creating resilience in people, process, and for our planet. You have read about the evolution of sustainability in Chapter 2 and how human capital has become a significant focus within the sustainability strategy. You will also have read about change management and how organizations sustain change – or become resilient.

In this final chapter, we will close by focusing on what it means to be resilient, through a few different case-in-point examples.

What Is It

As per Dr. Frederick Southwick wrote in Leadership Today, resilience can be defined "as the ability to regain balance following exposure to an adverse event or events." He further clarified, "resilience is not an end state of being, but rather a process of progressive growth within a risky landscape across the lifespan" (Southwick et al. 2016). Let us explore that definition through the lens of health and safety as we talk with Mr. Peter Evans, CSP, Deputy Director, Marine Corp Base Hawaii **s**afety **o**ffice. Mr. Evans had conducted research on resilience in health and systems safety for his master's thesis, striving to understand how near-misses, incidents, and accidents impact workers and their ability to remain focused on the mission (Evans and Marquette 2023).

People are the foundation of organizations, of communities. How leaders lead and workers work is essentially "the thread of resilience" that is woven through the organization and when an adverse event occurs, as

Connecting the Dots: Environment, Health, Safety, and Sustainability,
First Edition. Kathy A. Seabrook and Tanis J. Marquette.

Dr. Southwick discusses in Leadership and Resilience Chapter of Leadership Today, pulling that "resilience" thread straightens the organization. In speaking with Mr. Evans, through his research, he explains that thread is spun of training, procedures, rules, and repetition.

As Mr. Evans engaged with the Marines and civilians on base, the project team evaluated the Marines against a preset rubric to gauge effectiveness of the organizational structure and systems-thinking through the rigorous schedules. The Project team attempted to answer whether the Marines were resiliently and efficiently driving tasks to completion as they overcame each challenge. The key to the puzzle was the people and their ability to manage psychosocial stressors throughout the training and daily regimen of tasks to be done.

From Mr. Evan's research, we learn these elements cannot come together without sensemaking (Kilskar et al. 2020). From a resiliency lens, sensemaking is crucial for helping organizations and individuals navigate and recover from disruptions. It involves interpreting and understanding crises or unexpected events to restore order and develop adaptive responses. According to Weick (1993) in his study "The Collapse of Sensemaking in Organizations: The Mann Gulch Disaster,“ effective sensemaking allows for the reconstruction of meaning and guides actions during crises, enhancing organizational resilience. Sutcliffe and Vogus (2003) in "Organizing for Resilience" highlight that resilient organizations use sensemaking to quickly adapt, learn, and emerge stronger from adverse situations, ensuring long-term sustainability and robustness.

In Mr. Evans' case, he is striving to keep Marines focused on the resiliency of Marines. This parallels similar training activities earlier in the authors' career, the reason for emergency response, and confined space drills.

Another key aspect of people's resilience is the ability of people to withstand loneliness (Havens and Marquette 2023). The authors make that statement after having experienced COVID. We experienced first-hand our colleagues, family and friends struggle with being kept in isolation. Consider the long-term impacts of that combination of loneliness and COVID, then further consider, for some, the change to hybrid working. The 21st Surgeon General of the United States has been speaking for some time about the detrimental effects of loneliness (NPR 2020). Introduce the combination of COVID and hybrid-working and isolation doubled, people's mental health significantly suffered (HR Morning 2022). However, as COVID has ended, the statistics on loneliness have not improved. Mental health continues to be a significant challenge across our communities and workplaces globally with 7 out of 10 employers noting an increase in mental health issues in their workforce in 2023, up from 44% of employers in 2022, according to a recent Business Group on Health survey. (Health Payer Intelligence, 2023) People are suffering and have for some time; however, the convergence of the pandemic and hybrid-working elevated loneliness to a place where it is now significantly more visible, but alas not new (Holt-Lundstad et al. 2010). We raise awareness of loneliness here, unfortunately not because we have solutions, but because as Dr. Southwick contends, resilience is borne "following exposure" to adverse event however we have yet to realize loneliness-resilience.

Process

In the authors' estimation, a resilient process is one designed for repeatability. That is, a resilient process will continue beyond the initial implementation and champion. In Chapter 10, we discussed identifying champions in the organization to increase opportunities for early success and reducing the opportunities for nay-sayers to sabotage an initiative. Resilient processes are about creating sufficient consistency within a department or function so that if and when that champion moves on from that role, the excitement and success do not move on with them.

Initial implementation planning strives to locate processes and teams most accepting, if not in need of change, the most willing to adopt and ideally identify the early win-win to foster the example and spread positivity to adjacent teams and processes. As implementation continues and team leaders and embedded champions are promoted or take on more responsibility, mechanisms must be in place to ensure the program and corresponding processes continue and do not leave with the champions. Champions must share not only their passions but also how they execute. How the work is done must be documented and proliferated consistently throughout the team. Policies and procedures should updated (i.e. documented) and shared across the department and aligned across sites.

S = Specific
M = Measurable
A = Achievable
R = Relevant
T = Time-Bound
GOALS

Policy and procedures should be updated to reflect overarching corporate governance, goals, and targets. For example, new climate targets to reduce greenhouse gas emissions may be reflected in procurement by modifications to lightweight packaging specifications. Following our same procurement example, to further ensure resilience in our procurement processes, updating roles and responsibilities to reflect SMART goals that link back to the corporate goals could be written to reflect reducing packaging waste or logistics-related greenhouse gas emissions.

As discussed in the previous section, integrating elements of environmental stewardship, social responsibility, and governance into all functions' policies, procedures, training, and practices supports creating the culture of alignment and adaptability that enables the quick recovery in times of change when reinforced through regular communications and repetition. As Dr. Southwick contends, resilience is the thread that is woven throughout the department and in times of change pulled taut to bring strength to the team. Simply put, building resilience is about repeatability. To foster adoption, we can borrow from our marketing colleagues mantra of seven times, seven ways, meaning to best integrate change and engrain a new behavior or practice into the culture of an organization, we want to strive to communicate the expectations through multiple channels, for example, email, training, procedure updates, and multiple times to provide the content consistently (repeatedly) but through different methods to provide for the different way people receive information (Marquette and Windsor 2024).

And we all know the old adage – the more things change, the more they stay the same. Thus, the final mechanism to maintain a resilient process is a robust change management process that includes personnel changes, reassignments, and transition planning to help organizations track critical activities, thereby limiting the opportunity for activities to be "lost" with an employee departure.

Planet

And finally, let us turn our attention to the resilience of our planet. As environmental health safety and sustainability professionals, the opportunity is twofold. One to first introduce measures to reduce the health and safety impacts to workers in the near term while we switch to our sustainability hats and strive to reduce the impacts to the environment in the near to long term to ensure our environment remains inhabitable. Secondarily, if not primarily, not only do we need to be more resilient in our approach to preserving the natural environment when appropriate in our business decisions, but we also need to be regenerative wherever possible. Most organizations have turned to net zero pledges and decarbonization strategies to meet that commitment, according to the World Economic Forum (WEF 2021).

The impacts of climate change may still seem elusive to some; however, the aftermath of severe weather events creates changes in air quality, access to clean water, and infrastructure (American Public Health Association 2024). These outcomes, depending on the type of weather event, can lead to increases in:

- Increases in mental distress
- Increased allergy-related illnesses
- Dehydration and heat stroke risk
- Aggravation of cardiovascular illness
- Aggravation of respiratory illnesses
- Waterborne contamination and illnesses

See Figure 15.1 "Climate change impact and the EHS professional (Jones, Marquette 2023)."

Preparing workers for severe weather events through heat stress and vector-borne first aid training and annual emergency response exercises that include weather scenarios will set the stage for sensemaking foundational thinking (scenario-training) as discussed earlier in this section during our interview with Mr. Evans. Additional monitoring of the local community may also be required to support Environmental Justice Assessments related to safeguarding disadvantaged populations and anticipated future permitting applications to support enhanced infrastructure.

More substantially, organizations with heavy manufacturing or chemical operations may consider reassessing their containment walls to ensure sufficiency to withstand stormwaters and/or floodwaters should climate scenario analysis indicate the site and local community are within a 100-year flood plain. Ensuring workers have appropriate and safe shelter-in-place and evacuation stations, clear of electrical systems and chemical contamination, may require elevating equipment in some locations.

Organizations likely began a deeper review of their climate impacts with the emergence of the Task force on climate-related financial disclosures (TCFD) as discussed in Chapter 5. Site infrastructure and community access may be included in one of several different climate scenario analyses that are considered as part of a strong climate resilience and mitigation strategy. By capturing the climate risks and opportunities, strategies can be developed to not only mitigate the risk, but increase adaptability, that is, the resilience of the site.

Involving the organization, the community, and local emergency responders together with their insurance carrier in the development of mitigation strategies through stakeholder engagement increases the ability to return to a steady state sooner after a severe weather event.

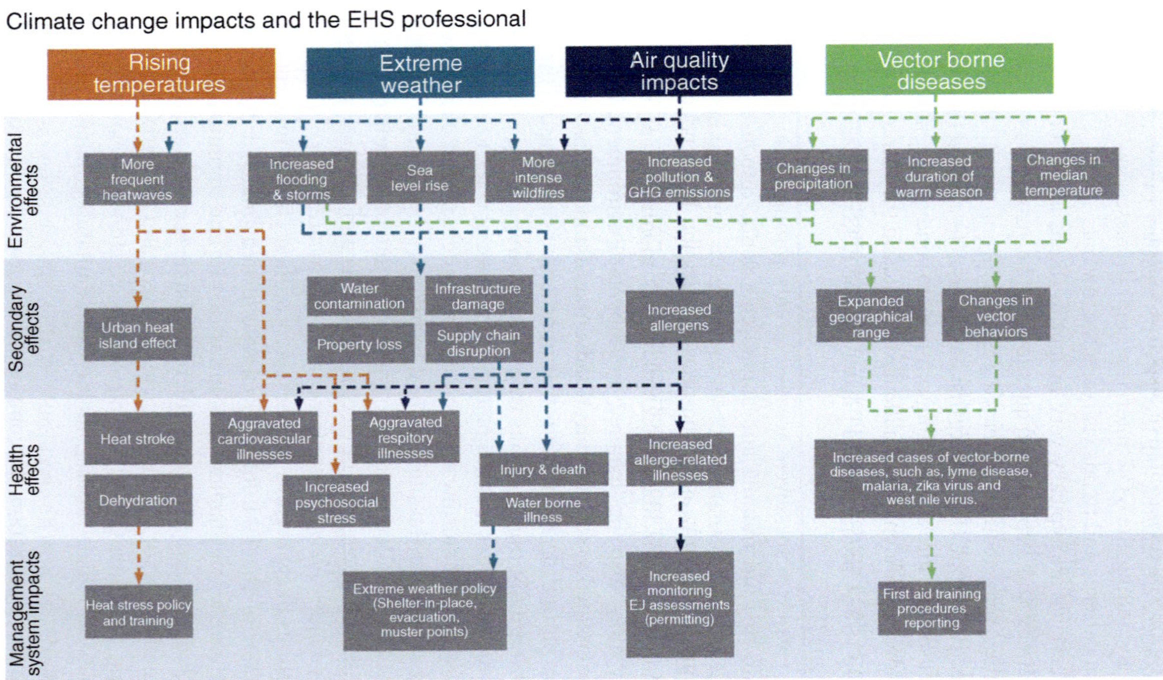

Adapted from material from JKGeography.com

Figure 15.1 Climate change and the EHS professional (2023).

To better understand how one global health-care organization was considering their carbon emissions, we spoke to Dr. Victoria Emerick, Executive Director and Global Head, Corporate Sustainability and Operations at Bristol Myers Squibs (BMS) (Emerick and Marquette 2024). Dr. Emerick shared that BMS's focus was squarely on the intersection of environment and health. She went on to explain that ~13 million people die each year from environmentally related health risk, and over 4 million people die prematurely every year from outdoor (ambient) air pollution and nearly the same from indoor air pollution. In 2020, 1 in 4 people lacked access to safe drinking water in their homes. She shared that the ILO predicts that under a 1.5 °C scenario, 2.2% of work hours will be lost by 2030, which is 80 million full-time jobs (ILO 2019). Carbon emissions and environmental impacts are viewed through the impacts on patients.

See Figure 15.2 "Moving beyond carbon at the intersection of environment and health (with Permission of Bristol Myers Squibb)."

Just as BMS has considered the interconnectedness of environment and health as they work to reduce their emissions, the interconnectedness of biodiversity and water stewardship has begun to unfold. Companies are increasingly focusing on the integrated strategies that address not only operational goals (energy, waste, and water), but positive biodiversity impacts as well (Reuters 2016). Sustainable development, as defined by the Brundtland report in 1987, is "development that meets the needs of the present without compromising the ability of future generations to meet their own needs." It sets the bar for future growth.

Nature-based solutions and the incorporation of species impact analysis in reporting has significantly increased. The United Nation's 15th session of the conference of parties (UN COP15) in 2022 produced the Global Biodiversity Framework (GBF), which outlined new goals for biodiversity to protect 30% of the world's land and oceans by 2030. The framework outlines 4 goals and 23 targets (Convention on Biological Diversity 2022). For the first time, the UN Climate Change Conference COP27 in 2023 included a day dedicated to the intersection of water and then an additional day dedicated to the intersection of biodiversity. Just as climate, water, and biodiversity are interconnected, so are environment, health, safety, and the sustainability professional.

This book has been focused on providing resources and tools to support you in creating the space within your organizations to help meet this ever-evolving challenge in your role, whether you are a dedicated environmental engineer, environmental, health and safety professional, or a sustainability leader. We are all part of a collaborative solution.

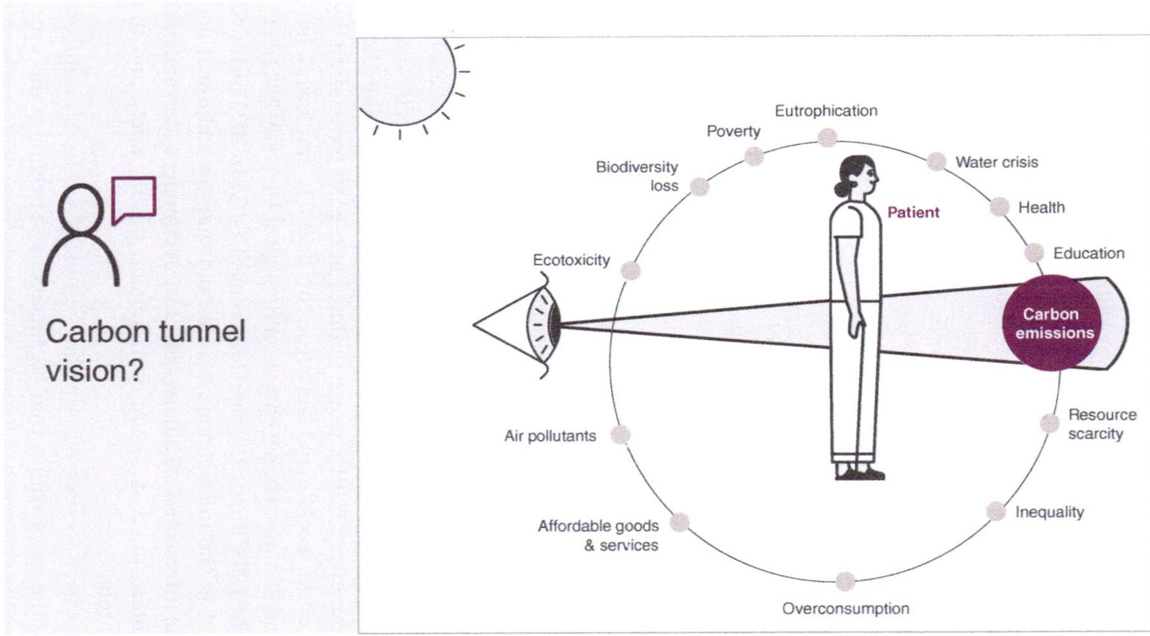

Figure 15.2 Moving beyond carbon at the intersection of environment and health. Source: Adapted from Moving beyond carbon tunnel vision with a sustainability data strategy, 2024.

References

American Public Health Association (2024). How climate change affects your health. https://www.apha.org/news-and-media/multimedia/infographics/how-climate-change-affects-your-health (accessed 29 June 2024).

Cognizant (2024). Moving beyond the carbon tunnel vision with a sustainability data strategy 8 February 2022. https://www.cognizant.com/us/en/insights/insights-blog/moving-beyond-carbon-tunnel-vision-with-a-sustainability-data-strategy-codex7121 (accessed 4 July 2024).

Convention on Biological Diversity (2022). COP15 final text of Kunming-Montreal global biodiversity framework. https://www.cbd.int/article/cop15-final-text-kunming-montreal-gbf-221222 (accessed 2 9 June 2024).

Emerick, V. and Marquette, T.J. (2024). 6 February, 2024 Interview with Victoria Emerick Executive Director and Global Head, Corporate Sustainability Strategy and Operations at Bristol-Myers Squibb. Emerick, Victoria (interviewee) and Marquette, Tanis J. (interviewer).

Evans, P. and Marquette, T.J. (2023). Interview with Peter Evans, Deputy Director, Marine Corps Base Hawaii Safety. Evans, Peter (interviewee) and Marquette, Tanis J. (interviewer).

Havens, J.C. and Marquette, T.J. (2023). Interview with John C. Havens, Director, Institute of Electrical and Electronics Engineers. Havens, John C. (interviewee) and Marquette, Tanis J. (interviewer).

Health Payer Intelligence (2023). Employers grapple with workforce mental health needs post-pandemic 23 August 2023. (accessed 15 June 2024).

Holt-Lundstad, J., Smith, T., and Layton, B. (2010). Social relationships and mortality risk: a meta-analytic review. *PLoS Medicine* 7 (7): e1000316. https://doi.org/10.1371/journal.pmed.1000316, pMC 2910600 PMID 20668659.

HR Morning (2022). How to foster connection, reduce loneliness in Hybrid or remote work 26 September 2022 (accessed 15 June 2024).

ILO (2019). Working on a warmer planet. https://www.ilo.org/sites/default/files/wcmsp5/groups/public/@dgreports/@dcomm/@publ/documents/publication/wcms:711919.pdf (accessed 4 July 2024).

Kilskar, S.S., Danielsen, B., and Johnsen, S.O. (2020). Sensemaking in critical situations and in relation to resilience – a review. *ASCE-ASME Journal of Risk and Uncertainty in Engineering Systems, Part B: Mechanical Engineering* 6 (1): 010801. https://doi.org/10.1115/1.4044789.

Marquette, T and Jones, M. (2024) The EHS Professionals' Compass for Navigating Climate Risk, Environmental Practice Specialty Webinar, July 13, 2023 hosted by American Society of Safety Professionals

Marquette, T. and Windsor, D. (2024). Operationalizing sustainability. https://youtu.be/gN24u4SNOsM?si=WE02pLyq_LtoVE5w 5 June, 2024 Webinar Hosted by Woodard & Curran

NPR (2020). In "together", former surgeon general write about importance of human connection 11 May, 2020. (accessed 15 June 2024).

Reuters (2016). More than 300 companies start reporting impact nature. https://www.reuters.com/sustainability/more-than-300-companies-start-reporting-impact-nature-2024-01-16 (accessed 29 June 2024).

Southwick, F.S., Martinit, B.L., Charney, D.S., and Southwick, S.M. (2016). Leadership and resilience. *Springer Texts in Business and Economics* 315–333. Https://doi.org/10.1007/978-3-319-31036-7_18.

Sutcliffe, K.M. and Vogus, T.J. (2003). Organizing for resilience. In: *Positive Organizational Scholarship: Foundations of a New Discipline* (ed. K.S. Cameron, J.E. Dutton, and R.E. Quinn), 94–110. Berrett-Koehler Publishers.

Weick, K.E. (1993). The collapse of sensemaking in organizations: the Mann Gulch disaster. *Administrative Science Quarterly* 38 (4): 628–652.

World Economic Forum (2021). From pledges to action: what's next for COP26 corporate commitments 1 December 2021. https://www.weforum.org/agenda/2021/12/from-pledges-to-action-what-s-next-for-cop26-corporate-commitments (accessed 30 June 2024).

Appendix

Figures by chapter and figure number

Connecting the Dots: Environment, Health, Safety, and Sustainability,
First Edition. Kathy A. Seabrook and Tanis J. Marquette.

The dimensions of sustainability dimensions: Figure 3.1.

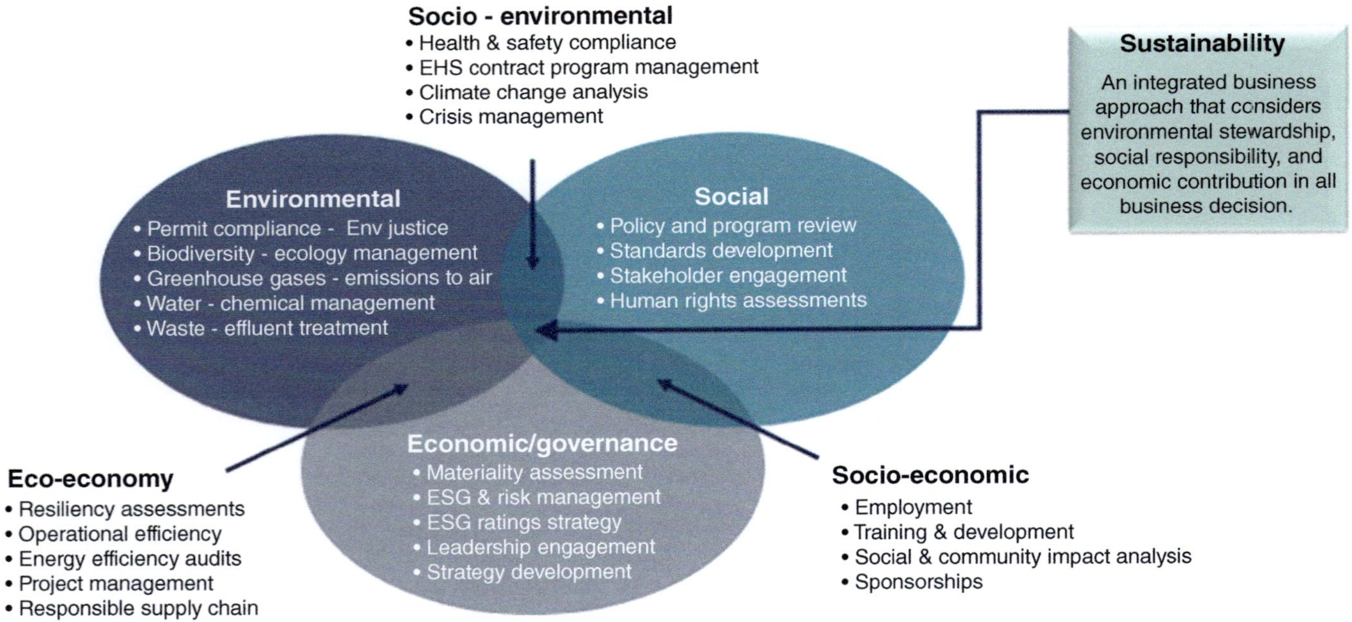

Sustainability/ESG issues: Figure 3.2.

Sustainability strategy and decision-making roadmap: Figures 4.1 and 9.2.

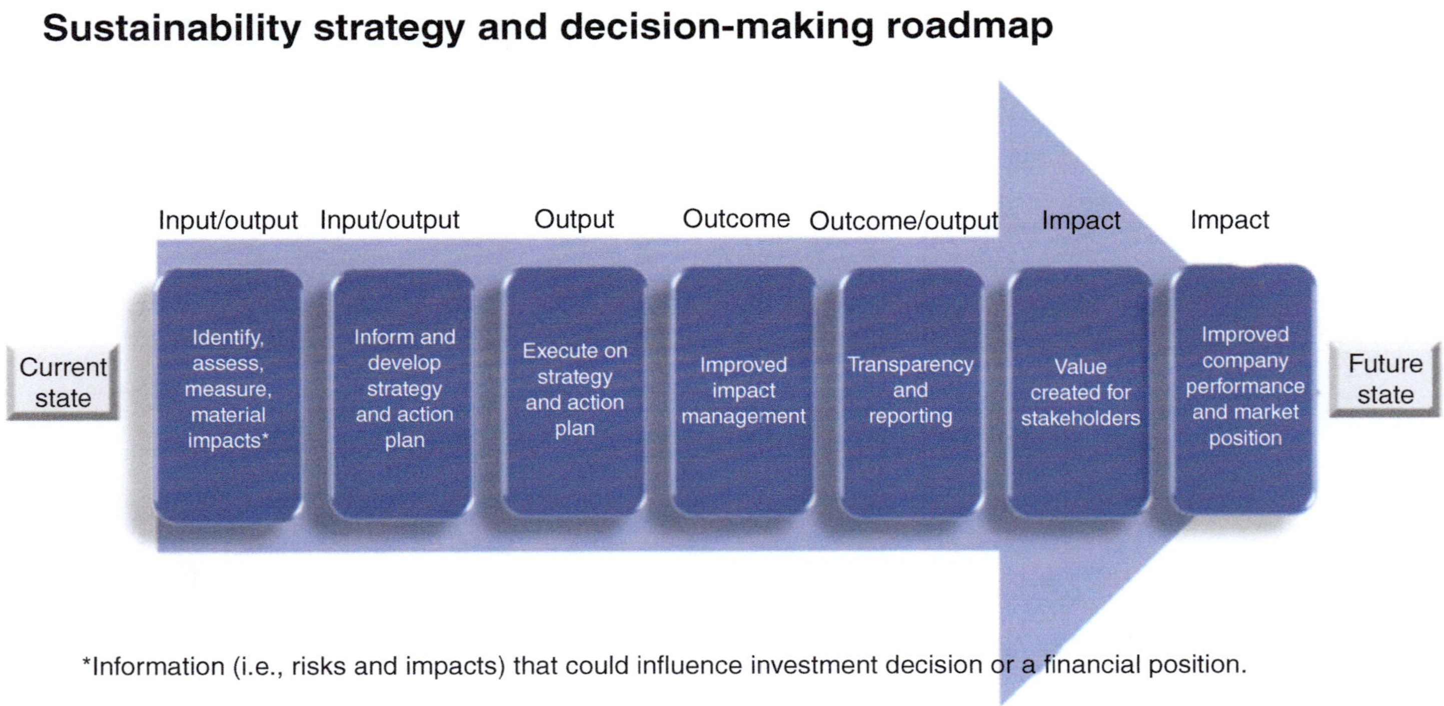

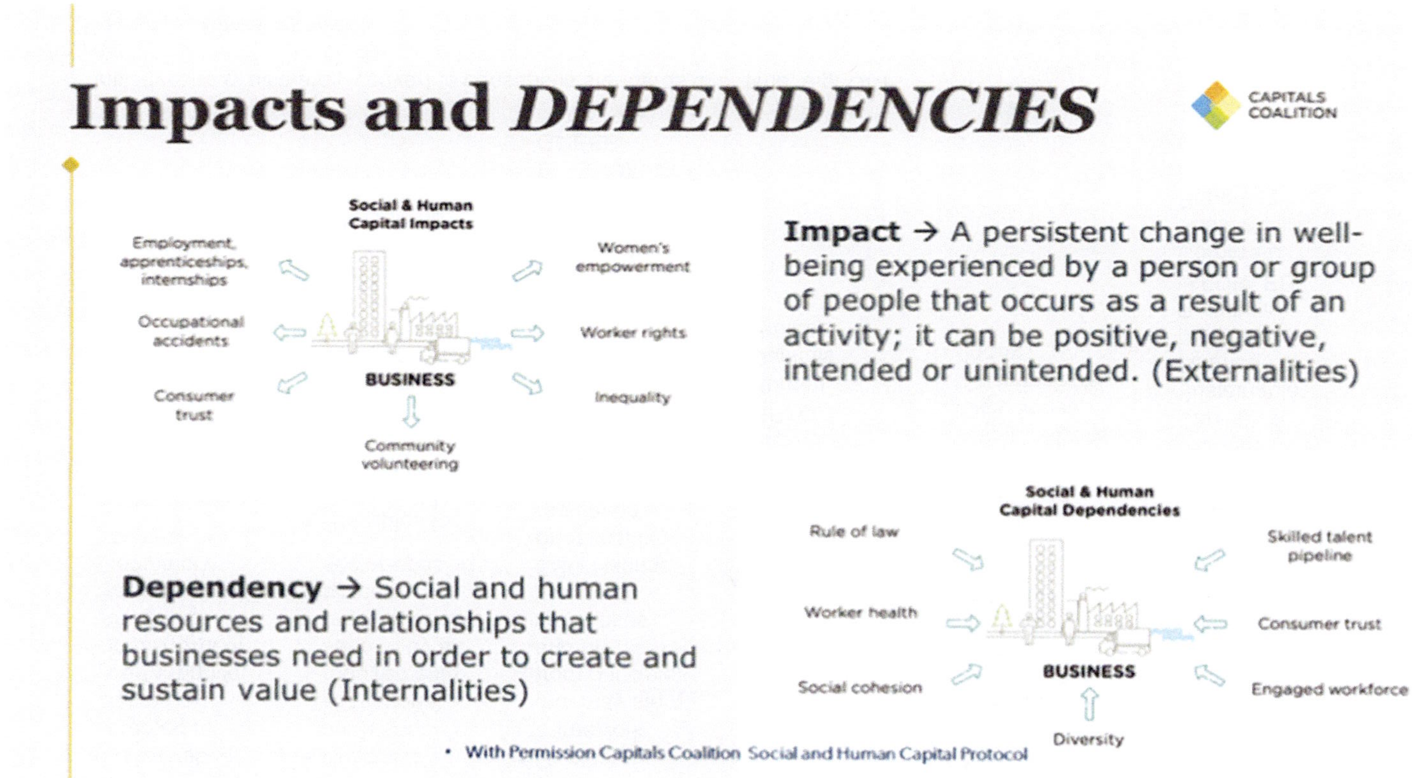

Figure 4.2 Impacts and dependencies: human and social capital example.

Initiatives*	Principles and declarations	Proprietary framework	Consensus standard	ESG Ratings & indices	Regulation
• Initiatives created to drive performance improvement in spite of lagging regulatory drivers • CDP (formerly known as Carbon Disclosure Project), Science-based Targets Initiative (SBTi), Net Zero Initiative (NZI)	• Principles are a set of aspirational requirements setforth by an organization, typically sector-specific • Principles for Responsible Investment (PRI), UN Global Compact, UN Blueprint for Business, Universal Declaration of Human Rights, UN Guiding Principles	• Disclosure framework that may stem from principles and/or consensus standards, ESG Ratings priorities, or regulation typically focused on an industry or sector • Responsible Care® Sustainability, Ecovadis (supply chain sustainability), USGBC LEED, BREEAM, GRESB	• Non-mandatory, usually consensus criteria for disclosing sustainability performance developed by industry professionals • ISO 45001, GRI 403, Human Capitals Protocol	• Survey-driven evaluations of sustainability performance for the financial industry (primarily investors) • MSCI, ISS, State Street, SustainAlytics, Dow Jones Sustainability Index	• Promulgated regulatory requirement to disclose certain sustainability criteria. May be topic-specific or broad. • US SEC S-K (Human Capital), Sustainable Finance Disclosures Regulation (SFDR),EU Corporate Sustainability Reporting Directive (CSRD)

*Initiatives may ultimately convert to consensus standards over time, e.g. CDP.

Figure 5.1 Framework types.

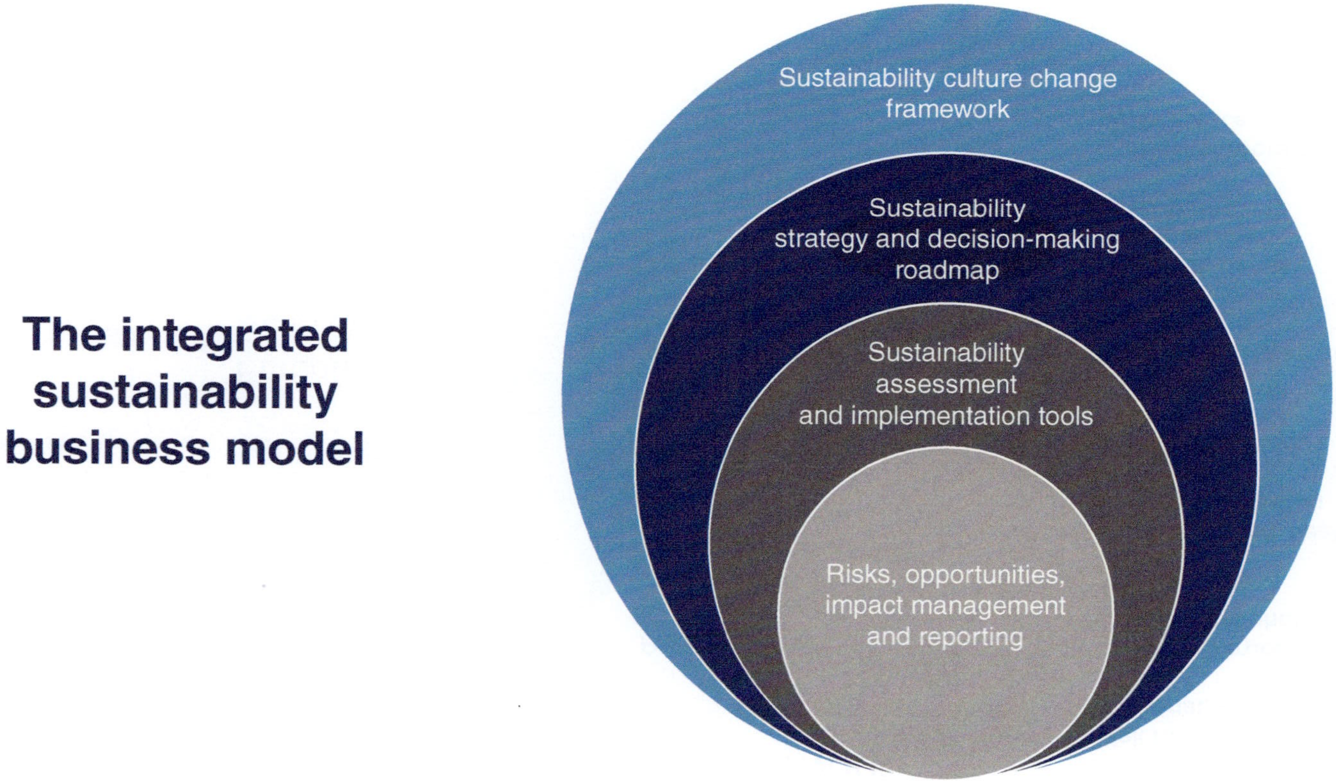

Figure 8.1 The integrated sustainability business model.

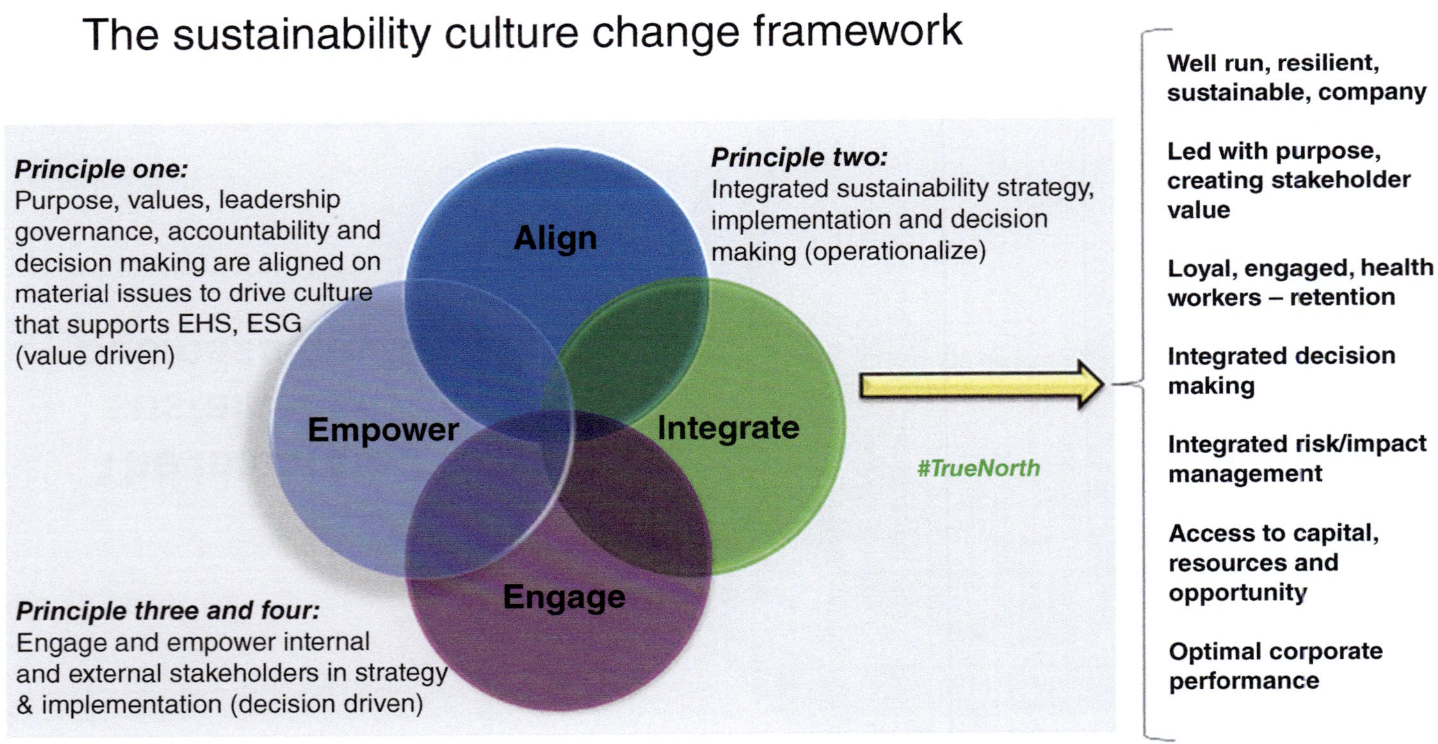

Figure 8.3 The sustainability culture change framework.

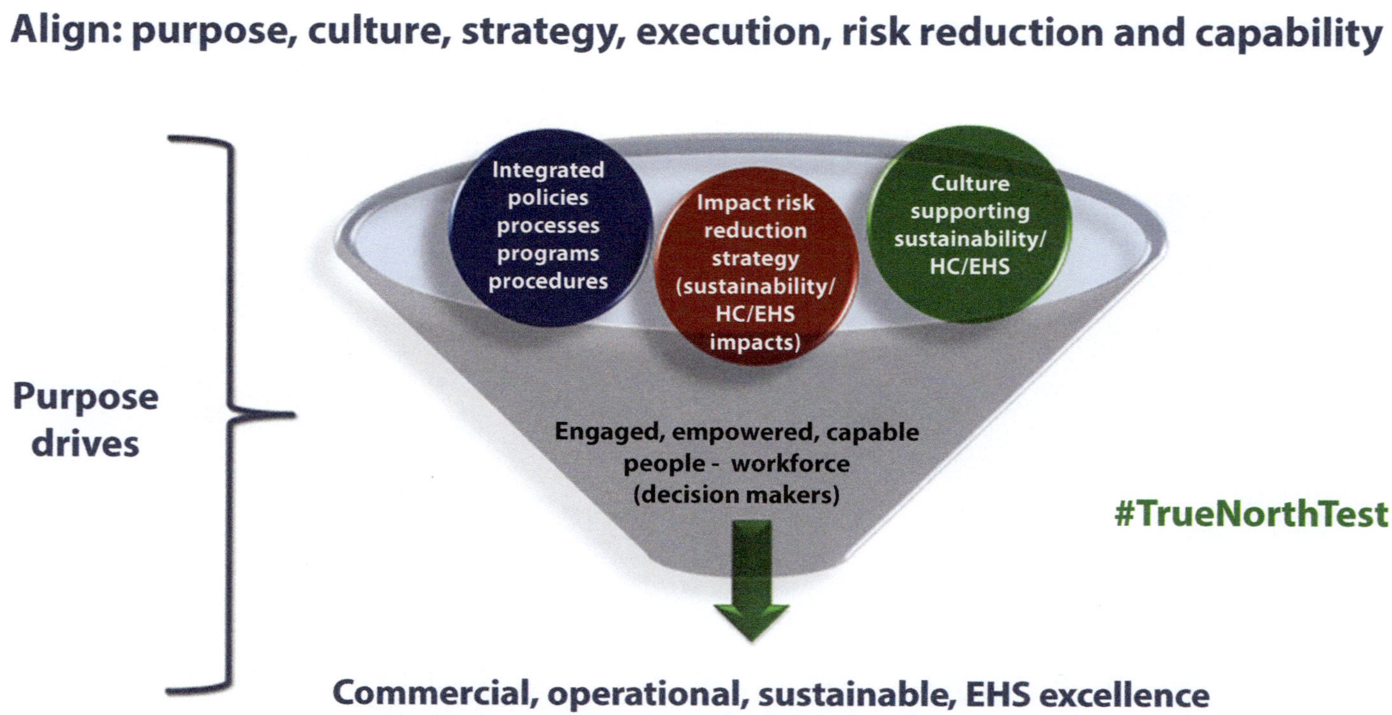

Figure 8.4 Aligning purpose, culture, strategy, execution, risk reduction and capability.

The Seabrook sustainability and ESG maturity curve: Figures 8.6 and 9.3.

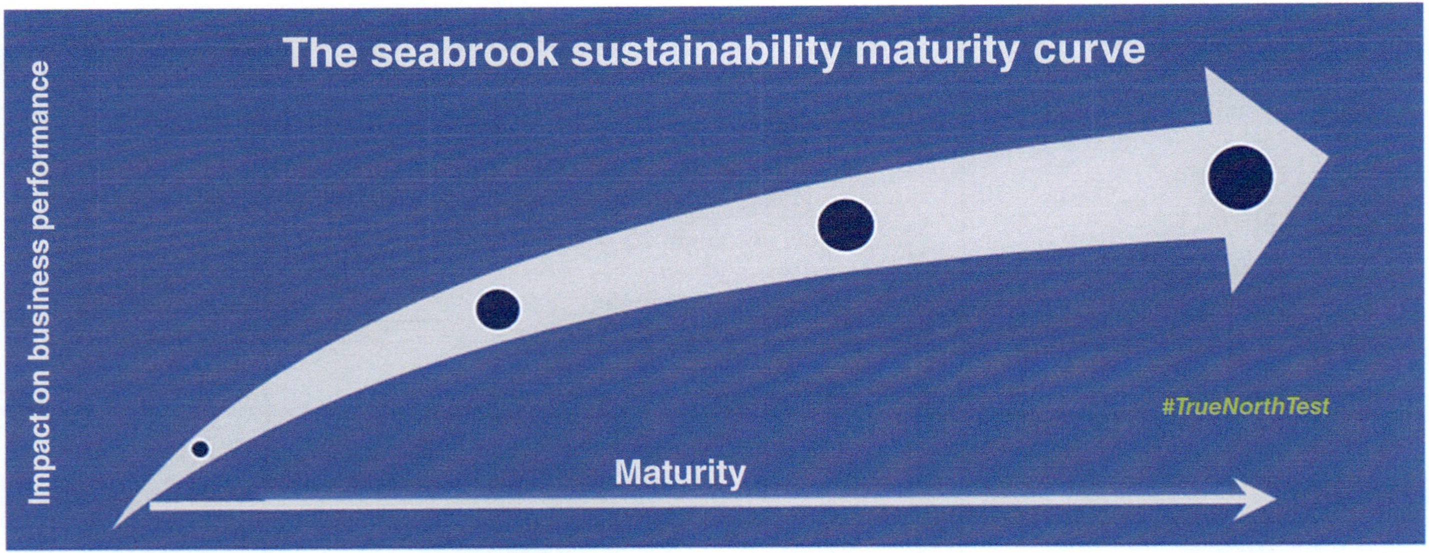

Reactive	Compliance	Managed	Integrated
No formal recognition, initiatives, strategy or leadership commitment to sustainability impact risk management. Responds to issues post event. No public disclosure.	Copliance with Stakeholder and regulatory requirements only. Focused on internal impact risks. Public disclosure as required.	Siloed approach to proactively identify, assess and manage internal and external impacts, risks and dependencies. No integrated role clarity, accountability systems or decision making. Some public disclosure.	Integrated, proactive, leadership led approach to strategy and decision-making. Identify and manage internal and external impacts, risks and dependencies. Integrated role clarity and accountability systems. Sector Leader. Public disclosure as competitive advantage. #TrueNorthTest

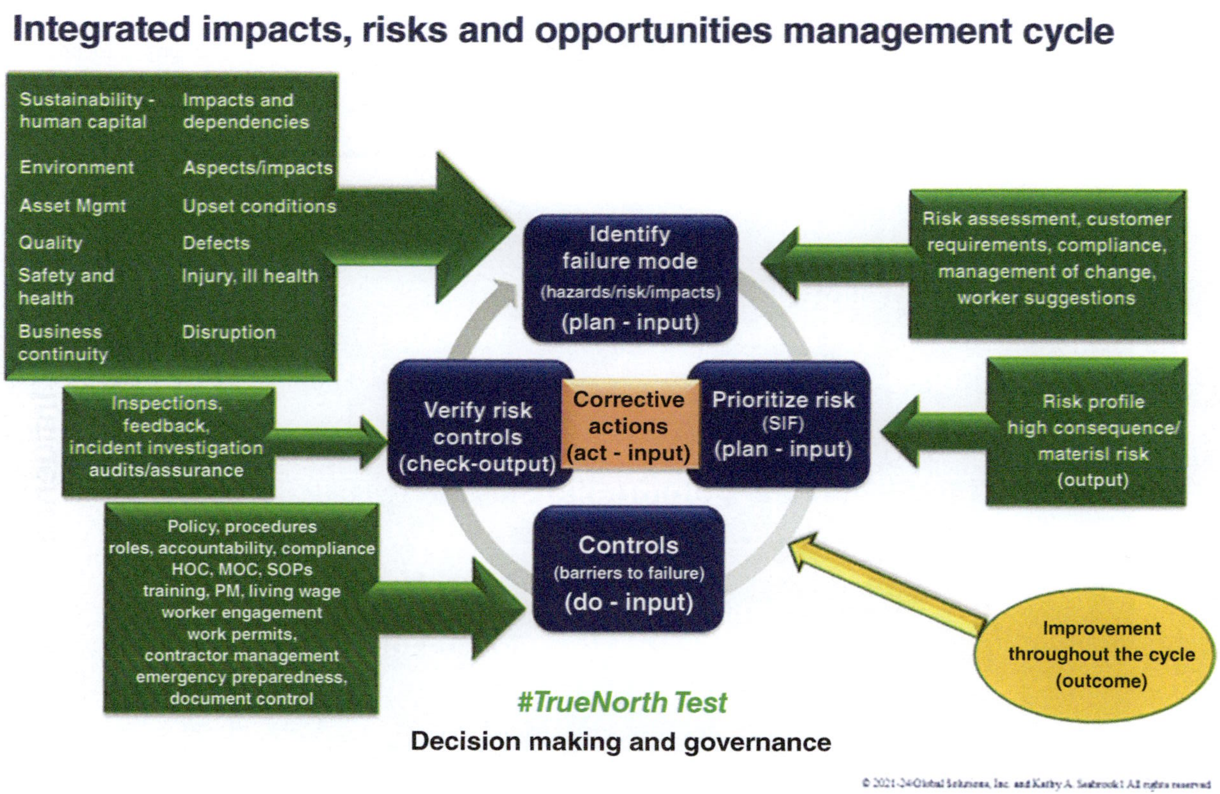

Figure 9.5 Integrated impacts, risks, and opportunities management cycle.

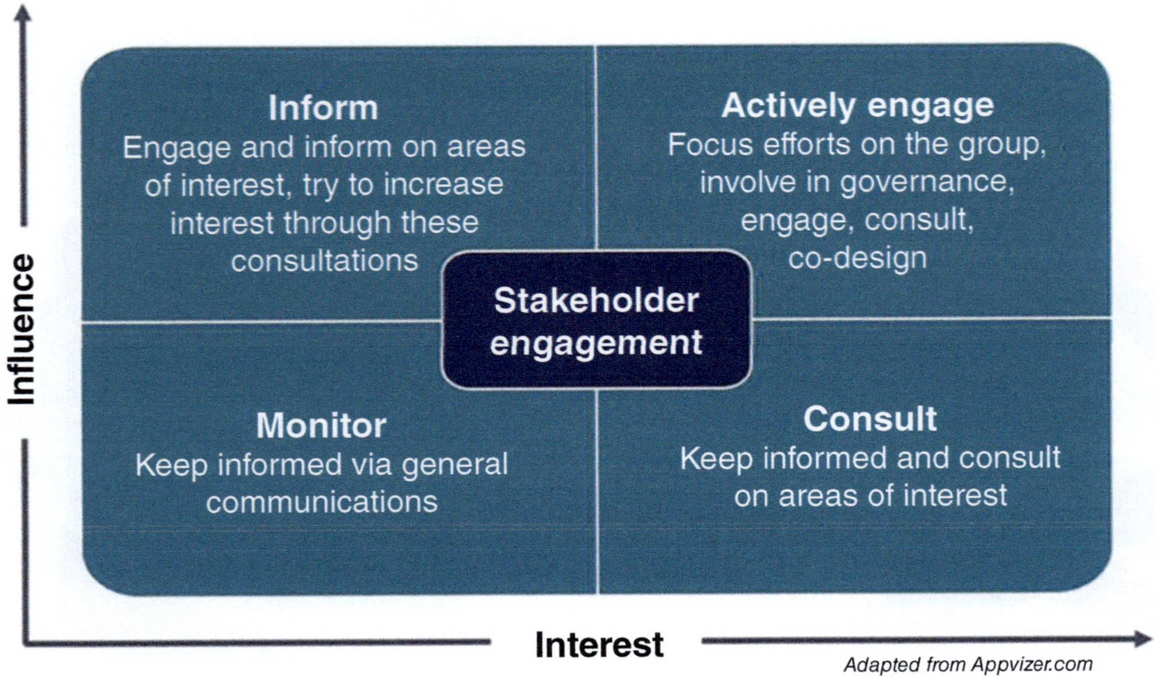

Figure 10.2 Stakeholder engagement matrix. Source: Marquette and Windsor (2024)/with permission of Woodard & Curran.

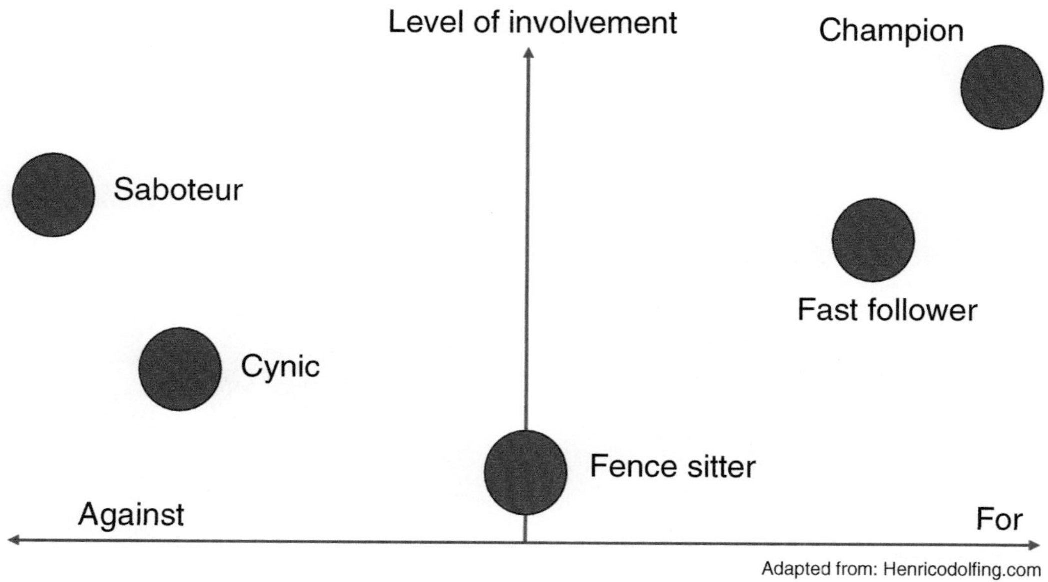

Figure 10.3 Stakeholder level of involvement. Source: Marquette and Windsor (2024)/with permission of Woodard & Curran.

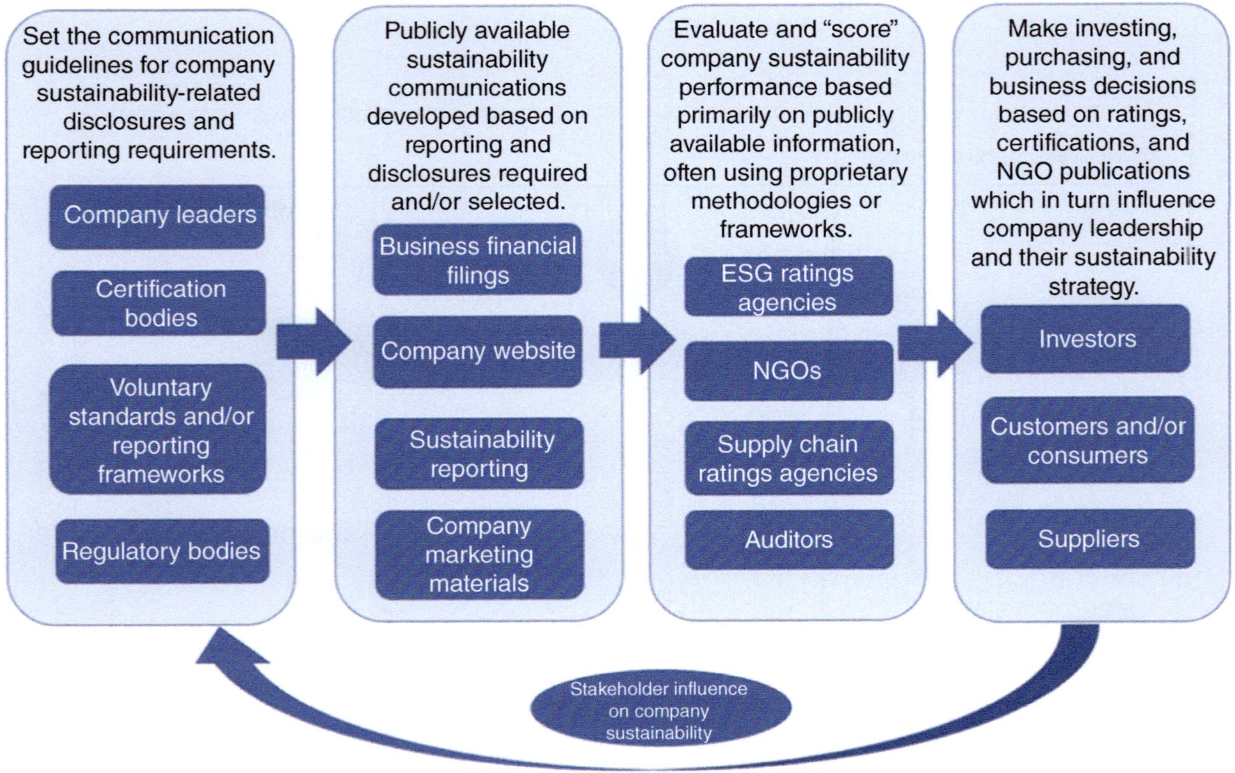

Figure 10.4 Stakeholder influence on sustainability strategy. Source: Marquette (2023)/with permission of American Society of Safety Professionals.

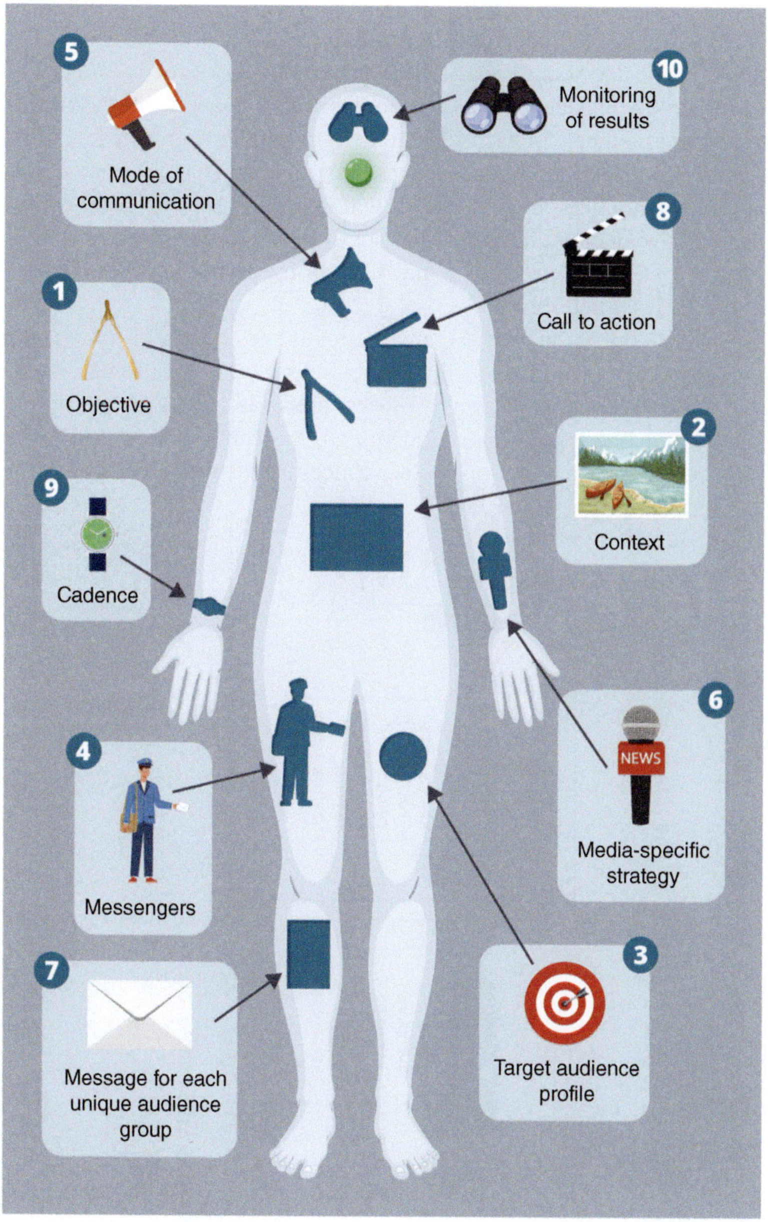

Figure 10.5 Communications plan sam. Source: Marquette and Windsor (2024)/ with permission of Woodard & Curran.

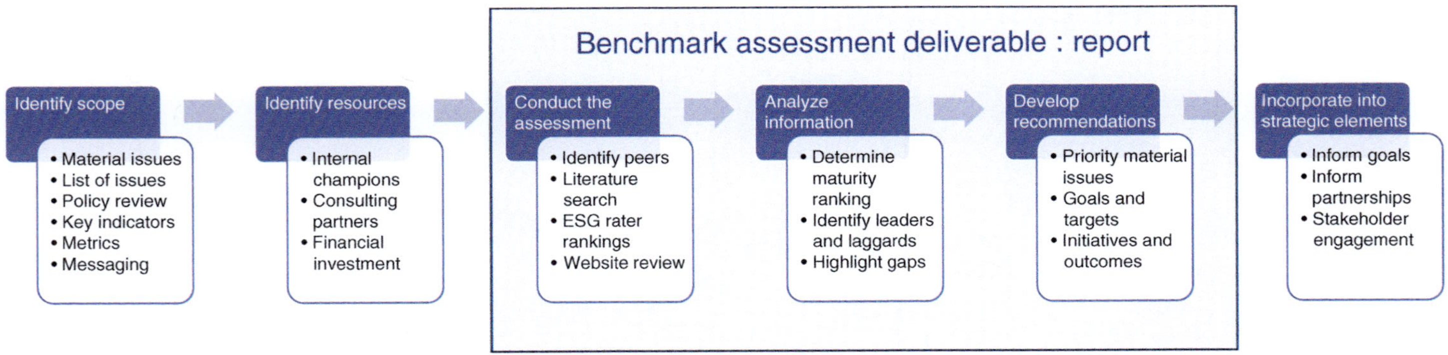

Figure 11.1 Benchmark assessment process.

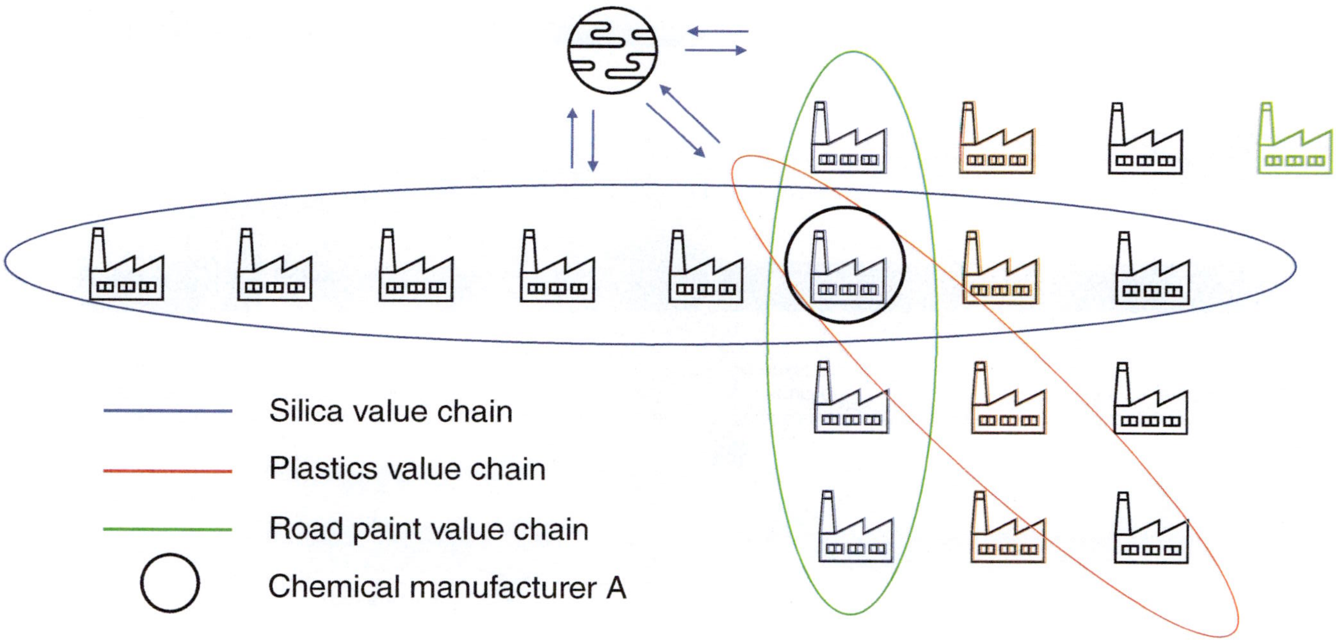

Figure 12.2 Mutual materiality.

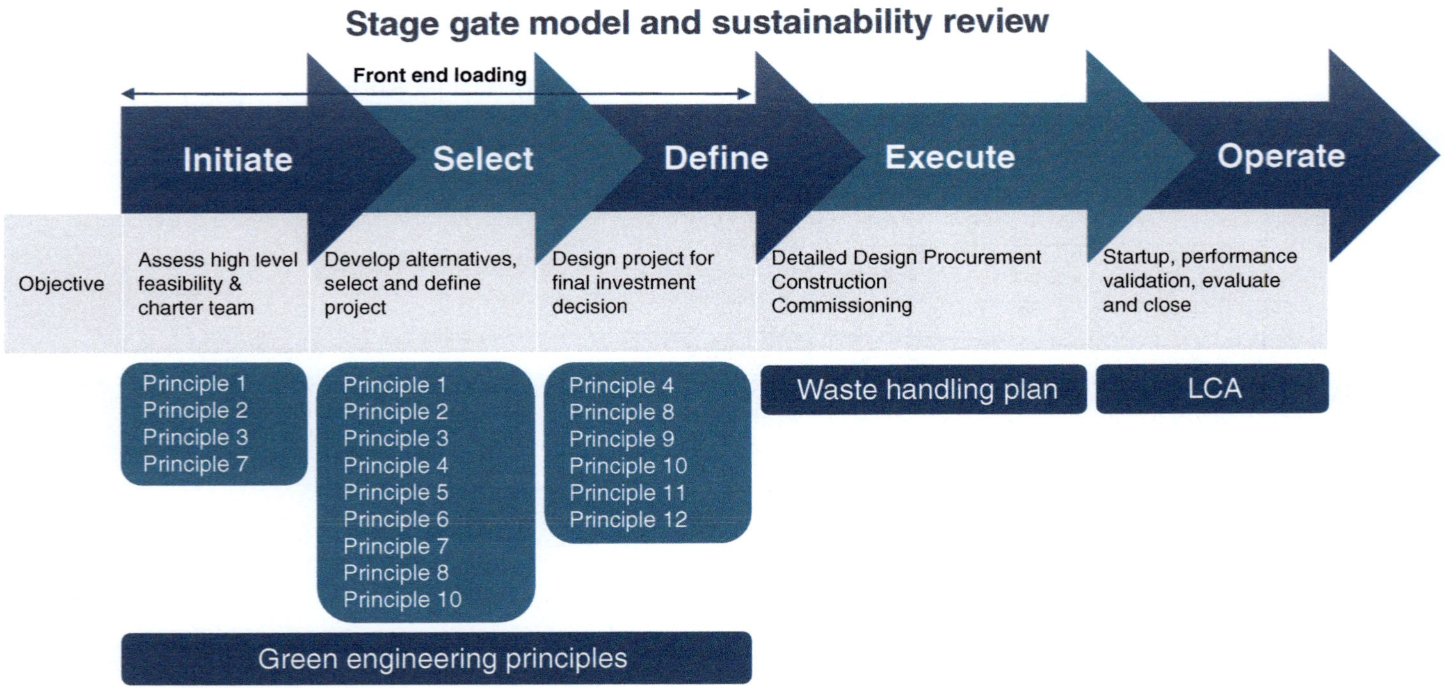

Figure 12.11 Engineering process incorporation of sustainability principles.

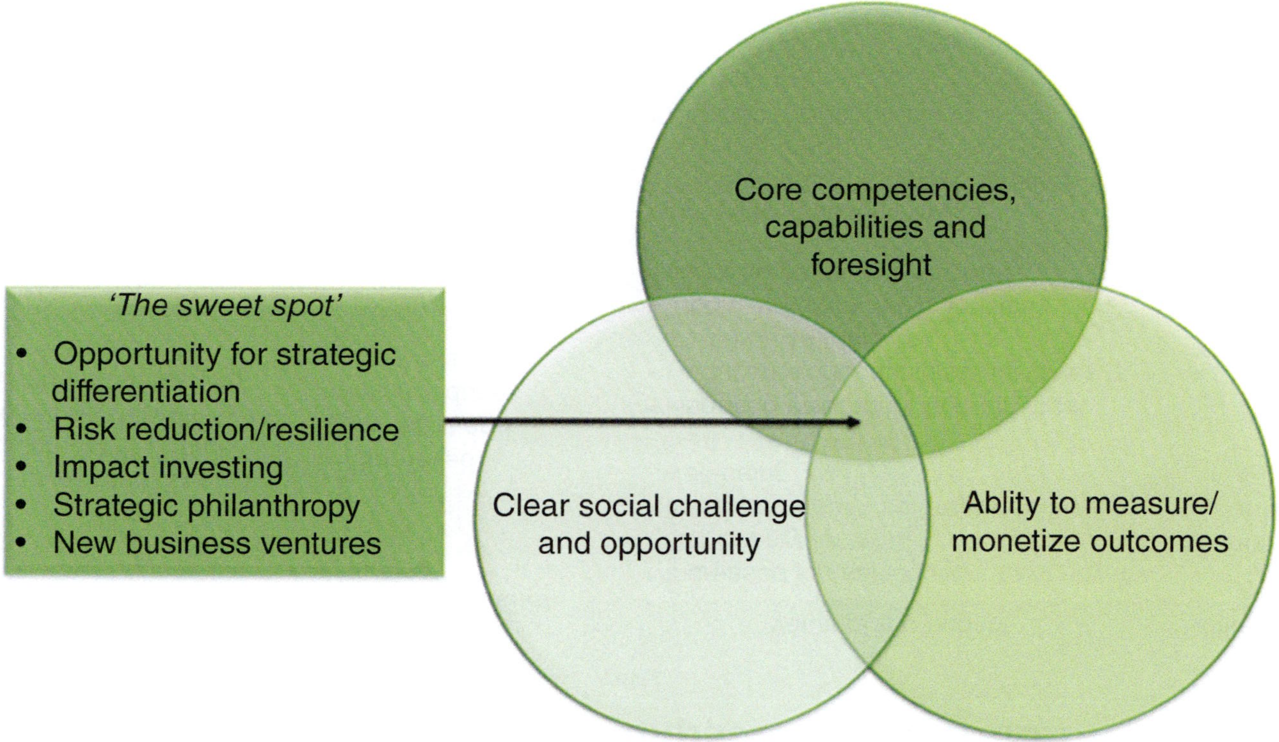

Figure 13.1 Value creation – "the sweet spot": philanthropy example.

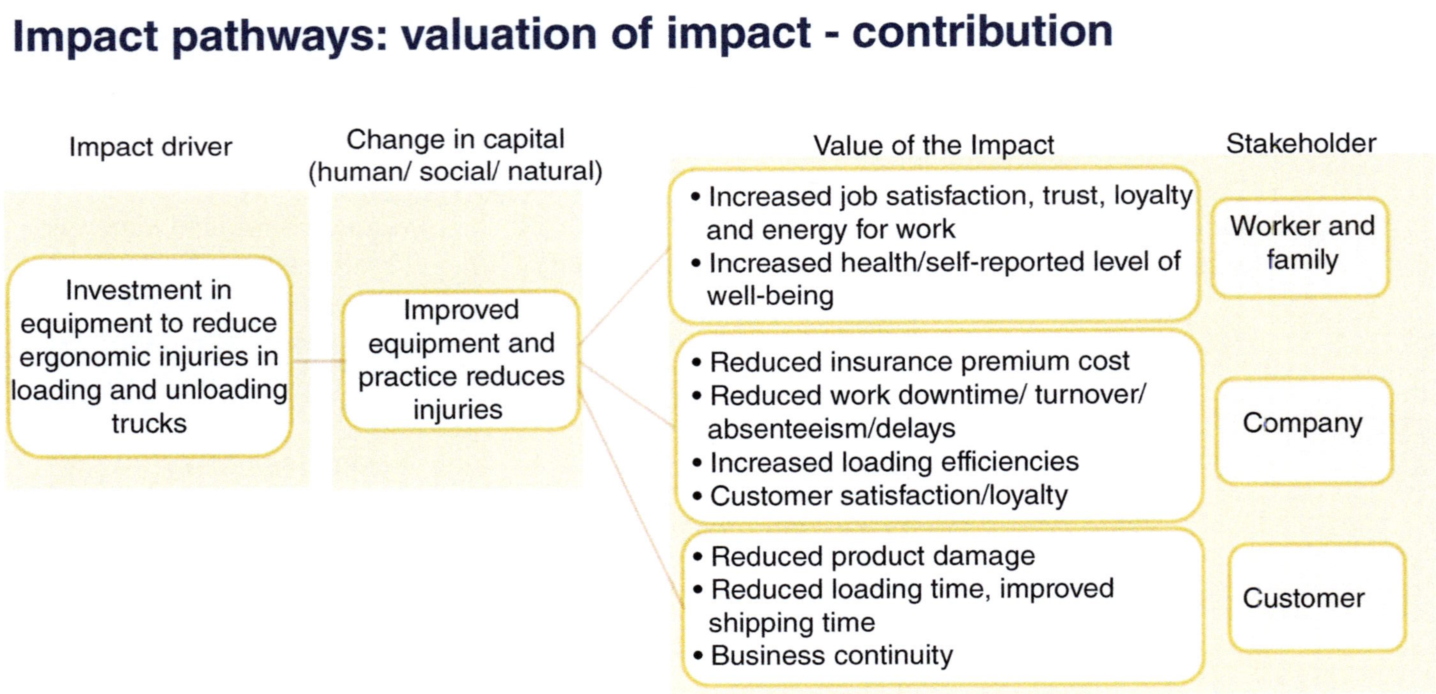

Figure 13.2 Impact pathways ergonomics example.

Index

Connecting the Dots: Environment, Health, Safety, and Sustainability,
First Edition. Kathy A. Seabrook and Tanis J. Marquette.

Printed and bound by CPI Group (UK) Ltd, Croydon, CR0 4YY
17/09/2026
14964310-0001